Groundwater Geochemistry

Broder J. Merkel · Britta Planer-Friedrich
Authors

Groundwater
Geochemistry

A Practical Guide to Modeling of Natural
and Contaminated Aquatic Systems

Edited by Darrell Kirk Nordstrom

2nd Edition

 Springer

Authors
Prof. Dr. Broder J. Merkel
TU Bergakademie Freiberg
Inst. Geologie
Gustav-Zeuner-Str. 12
09599 Freiberg
Germany
merkel@geo.tu-freiberg.de

Dr. Britta Planer-Friedrich
TU Bergakademie Freiberg
Inst. Geologie
Gustav-Zeuner-Str. 12
09599 Freiberg
Germany
b.planer-friedrich@geo.
tu-freiberg.de

Editor
Dr. Darrell K. Nordstrom
U.S. Geological Survey
3215 Marine St.
Boulder CO 80303
USA
dkn@usgs.gov

Additional material to this book can be downloaded from http://extras.springer.com.

ISBN 978-3-662-51750-5

ISBN 978-3-540-74668-3 (eBook)

DOI 10.1007/978-3-540-74668-3

Typesetting: Camera-ready by the Authors

Cover design: WMXDesign GmbH

Printed on acid-free paper

9 8 7 6 5 4 3 2 1

springer.com

Foreword

To understand hydrochemistry and to analyze natural as well as man-made impacts on aquatic systems, hydrogeochemical models have been used since the 1960's and more frequently in recent times.

Numerical groundwater flow, transport, and geochemical models are important tools besides classical deterministic and analytical approaches. Solving complex linear or non-linear systems of equations, commonly with hundreds of unknown parameters, is a routine task for a PC.

Modeling hydrogeochemical processes requires a detailed and accurate water analysis, as well as thermodynamic and kinetic data as input. Thermodynamic data, such as complex formation constants and solubility-products, are often provided as databases within the respective programs. However, the description of surface-controlled reactions (sorption, cation exchange, surface complexation) and kinetically controlled reactions requires additional input data.

Unlike groundwater flow and transport models, thermodynamic models, in principal, do not need any calibration. However, considering surface-controlled or kinetically controlled reaction models might be subject to calibration.

Typical problems for the application of geochemical models are:
- speciation
- determination of saturation indices
- adjustment of equilibria/disequilibria for minerals or gases
- mixing of different waters
- modeling the effects of temperature
- stoichiometric reactions (e.g. titration)
- reactions with solids, fluids, and gaseous phases (in open and closed systems)
- sorption (cation exchange, surface complexation)
- inverse modeling
- kinetically controlled reactions
- reactive transport

Hydrogeochemical models depend on the quality of the chemical analysis, the boundary conditions presumed by the program, theoretical concepts (e.g. calculation of activity coefficients) and the thermodynamic data. Therefore it is vital to check the results critically. For that, a basic knowledge about chemical and thermodynamic processes is required and will be outlined briefly in the following chapters on hydrogeochemical equilibrium (chapter 1.1), kinetics (chapter 1.2), and transport (chapter 1.3). Chapter 2 gives an overview on standard hydrogeochemical programs, problems and possible sources of error for modeling, and a detailed introduction to run the program PHREEQC, which is used in this book. With the help of examples, practical modeling applications are addressed and specialized theoretical knowledge is extended. Chapter 4 presents the results for the exercises of chapter 3. This book does not aim to replace a textbook but rather attempts to be a practical guide for beginners at modeling.

Table of Contents

1 Theoretical Background

1.1 Equilibrium reactions

1.1.1 Introduction

Chemical reactions determine occurrence, distribution, and behavior of aquatic species. Aquatic species are defined as organic and inorganic substances dissolved in water in contrast to colloids (1-1000 nm) and particles (> 1000 nm). This definition includes free anions and cations sensu strictu as well as complexes (chapter 1.1.5.1). The term complex applies to negatively charged species such as OH^-, HCO_3^-, CO_3^{2-}, SO_4^{2-}, NO_3^-, PO_4^{3-}, positively charged species such as $ZnOH^+$, $CaH_2PO_4^+$, $CaCl^+$, and zero-charged species such as $CaCO_3^0$, $FeSO_4^0$ or $NaHCO_3^0$ as well as organic ligands. Table 1 shows a selection of inorganic elements and examples of their dissolved species including both generally predominant and less common species.

Table 1   Selected inorganic elements and examples of aquatic species

Elements	
Major elements (>5 mg/L)	
Calcium (Ca)	Ca^{2+}, $CaOH^+$, CaF^+, $CaCl_2^0$, $CaCl^+$, $CaSO_4^0$, $CaHSO_4^+$, $CaNO_3^+$, $CaPO_4^-$, $CaHPO_4^0$, $CaH_2PO_4^+$, $CaP_2O_7^{2-}$, $CaCO_3^0$, $CaHCO_3^+$, $Ca_2(UO_2)(CO_3)_3^0$, $CaB(OH)_4^+$
Magnesium (Mg)	Mg^{2+}, $MgOH^+$, MgF^+, $MgSO_4^0$, $MgHSO_4^+$, $MgCO_3^0$, $MgHCO_3^+$
Sodium (Na)	Na^+, NaF^0, $NaSO_4^-$, $NaHPO_4^-$, $NaCO_3^-$, $NaHCO_3^0$, $NaCrO_4^-$
Potassium (K)	K^+, KSO_4^-, $KHPO_4^-$, $KCrO_4^-$
Carbon (C)	HCO_3^-, CO_3^{2-}, $CO_{2(g)}$, $CO_{2(aq)}$, $Ag(CO_3)_2^{2-}$, $AgCO_3^-$, $BaCO_3^0$, $BaHCO_3^+$, $CaCO_3^0$, $CaHCO_3^+$, $Ca_2(UO_2)(CO_3)_3^0$, $Cd(CO_3)_3^{4-}$, $CdHCO_3^+$, $CdCO_3^0$, $CuHCO_3^+$, $CuCO_3^0$, $Cu(CO_3)_2^{2-}$, $MgCO_3^0$, $MgHCO_3^+$, $MnHCO_3^+$, $NaCO_3^-$, $NaHCO_3^0$, $Pb(CO_3)_2^{2-}$, $PbCO_3^0$, $PbHCO_3^+$, $RaCO_3^0$, $RaHCO_3^+$, $SrCO_3^0$, $SrHCO_3^+$, $UO_2CO_3^0$, $UO_2(CO_3)_2^{2-}$, $UO_2(CO_3)_3^{4-}$, $Ca_2(UO_2)(CO_3)_3^0$, $ZnHCO_3^+$, $ZnCO_3^0$, $Zn(CO_3)_2^{2-}$
Sulfur (S)	SO_4^{2-}, SO_3^{2-}, $S_2O_3^{2-}$, S_x^-, $H_2S_{(g/aq)}$, HS^-, $Al(SO_4)_2^-$, $AlSO_4^+$, $BaSO_4^0$, $CaSO_4^0$, $CaHSO_4^+$, $Cd(SO_4)_2^{2-}$, $CdSO_4^0$, $CoSO_4^0$, $CoS_2O_3^0$, $CrO_3SO_4^{2-}$, $CrOHSO_4^0$, $CrSO_4^+$, $Cr_2(OH)_2(SO_4)_2^0$, $CuSO_4^0$, $Fe(SO_4)_2^-$, $FeSO_4^0$, $FeSO_4^+$, $HgSO_4^0$, $LiSO_4^-$, $MgSO_4^0$, $MgHSO_4^+$, $MnSO_4^0$, $NaSO_4^-$, $NiSO_4^0$, $Pb(SO_4)_2^{2-}$, $PbSO_4^0$, $RaSO_4^0$, $SrSO_4^0$, $Th(SO_4)_4^{4-}$, $Th(SO_4)_3^{2-}$, $Th(SO_4)_2^0$, $ThSO_4^{2+}$, $U(SO_4)_2^0$, USO_4^{2+}, $UO_2SO_4^0$, AsO_3S^{3-}, $AsO_2S_2^{3-}$, $AsOS_3^{3-}$, AsS_4^{3-}, $Cd(HS)_4^2$, $Cd(HS)_3^-$, $Cd(HS)_2^0$, $CdHS^+$, $Co(HS)_2^0$, $CoHS^+$, $Cu(S_4)_2^{3-}$, $Cu(HS)_3^-$, $Fe(HS)_3^-$, $Fe(HS)_2^0$, HgS_2^{2-}, $Hg(HS)_2^0$,

	$MoO_2S_2^{2-}$, $MoOS_3^{2-}$, $Pb(HS)_3^-$, $Pb(HS)_2^0$, $Sb_2S_4^{2-}$
Chlorine (Cl)	Cl^-, ClO^-, ClO_2^-, ClO_3^-, ClO_4^-, $AgCl_4^{3-}$, $AgCl_3^{2-}$, $AgCl_2^-$, $AgCl^0$, $BaCl^+$, $CaCl_2^0$, $CaCl^+$, $CdCl_3^-$, $CdCl_2^0$, $CdOHCl^0$, $CdCl^+$, $CoCl^+$, CrO_3Cl^-, $CrOHCl_2^0$, $CrCl_2^+$, $CrCl^{2+}$, $CuCl_3^{2-}$, $CuCl_4^{2-}$, $CuCl_3^-$, $CuCl_2^-$, $CuCl_2^0$, $CuCl^+$, $FeCl_3^0$, $FeCl_2^+$, $FeCl^{2+}$, $HgCl_4^{2-}$, $HgCl_3^-$, $HgCl_2^0$, $HgClI^0$, $HgClOH^0$, $HgCl^+$, $LiCl^0$, $MnCl_3^-$, $MnCl_2^0$, $MnCl^+$, $NiCl^+$, $PbCl_4^{2-}$, $PbCl_3^-$, $PbCl_2^0$, $PbCl^+$, $RaCl^+$, $ThCl_4^0$, $ThCl_3^+$, $ThCl_2^{2+}$, $ThCl^{3+}$, $TlCl_2^-$, $TlCl_4^-$, $TlCl^0$, $TlCl_3^0$, $TlCl_2^+$, $TlOHCl^+$, $TlBrCl^-$, $TlCl^{2+}$, UO_2Cl^+, UCl^{3+}, $ZnCl_4^{2-}$, $ZnCl_3^-$, $ZnCl_2^0$, $ZnOHCl^0$, $ZnCl^+$
Nitrogen (N)	NO_3^-, $AgNO_3^0$, $BaNO_3^-$, $CrNO_3^{2+}$, $CoNO_3^+$, $Hg(NO_3)_2^0$, $HgNO_3^+$, $Mn(NO_3)_2^0$, $Ni(NO_3)_2^0$, $NiNO_3^+$, $TlNO_3^{2+}$, NO_2^-, $NO_{(g/aq)}$, $NO_{2(g/aq)}$, $N_2O_{(g/aq)}$, $NH_{3(g/aq)}$, $HNO_{2(g/aq)}$, NH_4^+, $Cr(NH_3)_4(OH)_2^+$, $Cr(NH_3)_5OH^{2+}$, $Cr(NH_3)_6Br^{2+}$, $Cr(NH_3)_6^{3+}$, $HgNH_3^{2+}$, $Hg(NH_3)_2^{2+}$, $Hg(NH_3)_3^{2+}$, $Hg(NH_3)_4^{2+}$, $Ni(NH_3)_2^{2+}$, $Ni(NH_3)_6^{2+}$
Silicon (Si)	$H_4SiO_4^0$, $H_3SiO_4^-$, $H_2SiO_4^{2-}$, SiF_6^{2-}, $UO_2H_3SiO_4^+$
Minor elements (0.1-5 mg/L)	
Boron (B)	$B(OH)_3^0$, $BF_2(OH)_2^-$, BF_3OH^-, BF_4^-, $CaB(OH)_4^+$
Fluorine (F)	F^-, HF^0, HF_2^-, AgF^0, AsO_3F^{2-}, $HAsO_3F^-$, AlF_4^-, AlF_3^0, AlF_2^+, AlF^{2+}, $BF_2(OH)_2^-$, BF_3OH^-, BF_4^-, BaF^+, CaF^+, CdF_2^0, CdF^+, CrF^{2+}, CuF^+, FeF_3^0, FeF^+, FeF_2^+, FeF^{2+}, MgF^+, MnF^+, NaF^0, PO_3F^{2-}, HPO_3F^-, $H_2PO_3F^0$, PbF_4^{2-}, PbF_3^-, PbF_2^0, PbF^+, $SbOF^0$, $Sb(OH)_2F^0$, SiF_6^{2-}, SnF_3^-, SnF_2^0, SnF^+, SrF^+, ThF_4^0, ThF_3^+, ThF_2^{2+}, ThF^{3+}, $UO_2F_4^{2-}$, UF_6^{2-}, $UO_2F_3^-$, UF_5^-, UF_4^0, $UO_2F_2^0$, UO_2F^+, UF_3^+, UF_2^{2+}, UF^{3+}, ZnF^+
Iron (Fe)	Fe^{2+}, Fe^{3+}, $Fe(OH)_3^-$, $Fe(OH)_2^0$, $FeOH^{2+}$, $Fe(OH)_2^+$, $Fe(OH)_3^0$, $Fe(OH)_4^-$, $Fe_2(OH)_2^{4+}$, $Fe_3(OH)_4^{5+}$, $FeCl_3^0$, $FeCl_2^+$, $FeCl^{2+}$, FeF^+, FeF^{2+}, FeF_2^+, FeF_3^0, $FeSO_4^0$, $Fe(SO_4)_2^-$, $FeSO_4^+$, $Fe(HS)_2^0$, $Fe(HS)_3^-$, $FePO_4^-$, $FeHPO_4^0$, $FeH_2PO_4^+$, $FeH_2PO_4^{2+}$
Strontium (Sr)	Sr^{2+}, $SrOH^+$, $SrSO_4^0$, $SrCO_3^0$, $SrHCO_3^+$
Trace elements (<0.1 mg/L)	
Lithium (Li)	Li^+, $LiOH^0$, $LiCl^0$, $LiSO_4^-$
Beryllium (Be)	Be^{2+}, BeO_2^{2-}, $BeSO_4^0$, $BeCO_3^0$
Aluminum (Al)	Al^{3+}, $AlOH^{2+}$, $Al(OH)_2^+$, $Al(OH)_3^0$, $Al(OH)_4^-$, AlF^{2+}, AlF_2^+, AlF_3^0, AlF_4^-, $AlSO_4^+$, $Al(SO_4)_2^-$
Phosphorus (P)	PO_4^{3-}, HPO_4^{2-}, $H_2PO_4^-$, $H_3PO_4^0$, $CaPO_4^-$, $CaHPO_4^0$, $CaH_2PO_4^+$, $CaP_2O_7^{2-}$, $CrH_2PO_4^{2+}$, $CrO_3H_2PO_4^-$, $CrO_3HPO_4^{2-}$, $H_2PO_3F^0$, HPO_3F^-, PO_3F^{2-}, $FePO_4^-$, $FeHPO_4^0$, $FeH_2PO_4^+$, $FeH_2PO_4^{2+}$, $KHPO_4^-$, $MgPO_4^-$, $MgHPO_4^0$, $MgH_2PO_4^+$, $NaHPO_4^-$, $NiHP_2O_7^-$, $NiP_2O_7^{2-}$, $ThH_2PO_4^{3+}$, $ThH_3PO_4^{4+}$, $ThHPO_4^{2+}$, $UHPO_4^{2+}$, $U(HPO_4)_2^0$, $U(HPO_4)_3^{2-}$, $U(HPO_4)_4^{4-}$, $UO_2HPO_4^0$, $UO_2(HPO_4)_2^{2-}$, $UO_2H_2PO_4^+$, $UO_2(H_2PO_4)_2^0$, $UO_2(H_2PO_4)_3^-$
Chromium (Cr)	Cr^{3+}, $Cr(OH)^{2+}$, $Cr(OH)_2^+$, $Cr(OH)_3^0$, $Cr(OH)_4^-$, CrO_2^-, CrO_4^{2-}, $HCrO_4^-$, $H_2CrO_4^0$, $Cr_2O_7^{2-}$, CrF^{2+}, $CrCl^{2+}$, $CrCl_2^+$, $CrOHCl_2^0$, CrO_3Cl^-, $CrBr^{2+}$, CrI^{2+}, $CrSO_4^+$, $CrOHSO_4^0$, $Cr_2(OH)_2(SO_4)_2^0$, $CrH_2PO_4^{2+}$, $CrO_3H_2PO_4$, $CrO_3HPO_4^{2-}$, $Cr(NH_3)_6^{3+}$, $Cr(NH_3)_5OH^{2+}$, $Cr(NH_3)_4(OH)_2^+$, $Cr(NH_3)_6Br^{2+}$, $CrNO_3^{2+}$, $CrO_3SO_4^{2-}$, $KCrO_4^-$, $NaCrO_4^-$
Manganese (Mn)	Mn^{2+}, $MnOH^+$, $Mn(OH)_3^-$, MnF^+, $MnCl^+$, $MnCl_2^0$, $MnCl_3^-$, $MnSO_4^0$, $MnSe^0$, $MnSeO_4^0$, $Mn(NO_3)_2^0$, $MnHCO_3^+$
Cobalt (Co)	Co^{3+}, $Co(OH)_2^0$, $Co(OH)_4^-$, $Co_4(OH)_4^{4+}$, $Co_2(OH)_3^+$, $CoCl^+$, $CoBr_2^0$, CoI_2^0, $CoSO_4^0$, $CoS_2O_3^0$, $CoHS^+$, $Co(HS)_2^0$, $CoSeO_4^0$, $CoNO_3^+$
Nickel (Ni)	Ni^{2+}, $Ni(OH)_2^0$, $Ni(OH)_3^-$, Ni_2OH^{3+}, $Ni_4(OH)_4^{4+}$, $NiCl^+$, $NiBr^+$, $NiSO_4^0$,

	$NiSeO_4^0$, $NiHP_2O_7^-$, $NiP_2O_7^{2-}$, $Ni(NH_3)_2^{2+}$, $Ni(NH_3)_6^{2+}$, $Ni(NO_3)_2^0$, $NiNO_3^+$
Silver (Ag)	Ag^+, AgF^0, $AgCl^0$, $AgCl_2^-$, $AgCl_3^{2-}$, $AgCl_4^{3-}$, $AgBr^0$, $AgBr_2^-$, $AgBr_3^{2-}$, $AgSeO_3^-$, $Ag(SeO_3)_2^{3-}$, $AgNO_3^0$, $Ag(CO_3)_2^{2-}$, $AgCO_3^-$
Copper (Cu)	Cu^+, Cu^{2+}, $CuOH^+$, $Cu(OH)_2^0$, $Cu(OH)_3^-$, $Cu(OH)_4^{2-}$, $Cu_2(OH)_2^{2+}$, CuF^+, $CuCl^+$, $CuCl_2^0$, $CuCl_3^-$, $CuCl_4^{2-}$, $CuCl_2^-$, $CuCl_3^{2-}$, $CuSO_4^0$, $Cu(HS)_3^-$, $Cu(S_4)_2^{3-}$, $CuCO_3^0$, $Cu(CO_3)_2^{2-}$, $CuHCO_3^+$
Zinc (Zn)	Zn^{2+}, $ZnOH^+$, $Zn(OH)_2^0$, $Zn(OH)_3^-$, $Zn(OH)_4^{2-}$, ZnF^+, $ZnCl^+$, $ZnCl_2^0$, $ZnCl_3^-$, $ZnCl_4^{2-}$, $ZnOHCl^0$, $ZnBr^+$, $ZnBr_2^0$, ZnI^+, ZnI_2^0, $ZnSO_4^0$, $Zn(SO_4)_2^{2-}$, $Zn(HS)_2^0$, $Zn(HS)_3^-$, $ZnSeO_4^0$, $Zn(SeO_4)_2^{2-}$, $ZnHCO_3^+$, $ZnCO_3^0$, $Zn(CO_3)_2^{2-}$
Arsenic (As)	$H_3AsO_3^0$, $H_2AsO_3^-$, $HAsO_3^{2-}$, AsO_3^{3-}, $H_4AsO_3^+$, $H_2AsO_4^-$, $HAsO_4^{2-}$, AsO_4^{3-}, AsO_3S^{3-}, $AsO_2S_2^{3-}$, $AsOS_3^{3-}$, AsS_4^{3-}, AsO_3F^{2-}, $HAsO_3F^-$, $UO_2H_2AsO_4^+$, $UO_2HAsO_4^0$, $UO_2(H_2AsO_4)_2^0$
Selenium (Se)	Se^{2-}, HSe^-, H_2Se^0, $HSeO_3^-$, SeO_3^{2-}, $H_2SeO_3^0$, SeO_4^{2-}, $HSeO_4^-$, Ag_2Se^0, $AgOH(Se)_2^{4-}$, $FeHSeO_3^{2+}$, $AgSeO_3^-$, $Ag(SeO_3)_2^{3-}$, $Cd(SeO_3)_2^{2-}$, $CdSeO_4^0$, $CoSeO_4^0$, $MnSe^0$, $MnSeO_4^0$, $NiSeO_4^0$, $ZnSeO_4^0$, $Zn(SeO_4)_2^{2-}$
Bromine (Br)	Br^-, Br^{3-}, Br_2, BrO^-, BrO_3^-, BrO_4^-, $AgBr^0$, $AgBr_2^-$, $AgBr_3^{2-}$, $BaB(OH)_4^+$, $CdBr^+$, $CdBr_2^0$, $CoBr_2^0$, $CrBr^{2+}$, $PbBr^+$, $PbBr_2^0$, $NiBr^+$, $ZnBr^+$, $ZnBr_2^0$
Molybdenum (Mo)	Mo^{6+}, $H_2MoO_4^0$, $HMoO_4^-$, MoO_4^{2-}, $Mo(OH)_6^0$, $MoO(OH)_5^-$, MoO_2^{2+}, $MoO_2S_2^{2-}$, $MoOS_3^{2-}$
Cadmium (Cd)	Cd^{2+}, $CdOH^+$, $Cd(OH)_2^0$, $Cd(OH)_3^-$, $Cd(OH)_4^{2-}$, Cd_2OH^{3+}, CdF^+, CdF_2^0, $CdCl^+$, $CdCl_2^0$, $CdCl_3^-$, $CdOHCl^0$, $CdBr^+$, $CdBr_2^0$, CdI^+, CdI_2^0, $CdSO_4^0$, $Cd(SO_4)_2^{2-}$, $CdHS^+$, $Cd(HS)_2^0$, $Cd(HS)_3^-$, $Cd(HS)_4^{2-}$, $CdSeO_4^0$, $CdNO_3^+$, $Cd(CO_3)_3^{3-}$, $CdHCO_3^+$, $CdCO_3^0$
Antimony (Sb)	$Sb(OH)_3^0$, $HSbO_2^0$, SbO^+, SbO_2^-, $Sb(OH)_2^+$, $Sb(OH)_6^-$, SbO_3^-, SbO_2^+, $Sb(OH)_4^-$, $SbOF^0$, $Sb(OH)_2F^0$, $Sb_2S_4^{2-}$
Barium (Ba)	Ba^{2+}, $BaOH^+$, $BaCO_3^0$, $BaHCO_3^+$, $BaNO_3^-$, BaF^+, $BaCl^+$, $BaSO_4^0$, $BaB(OH)_4^+$
Mercury (Hg)	Hg^{2+}, $Hg(OH)_2^0$, $HgOH^+$, $Hg(OH)_3^-$, HgF^+, $HgCl^+$, $HgCl_2^0$, $HgCl_3^-$, $HgCl_4^{2-}$, $HgClI^0$, $HgClOH^0$, $HgBr^+$, $HgBr_2^0$, $HgBr_3^-$, $HgBr_4^{2-}$, $HgBrCl^0$, $HgBrI^0$, $HgBrI_3^{2-}$, $HgBr_2I_2^{2-}$, $HgBr_3I^{2-}$, $HgBrOH^0$, HgI^+, HgI_2^0, HgI_3^-, HgI_4^{2-}, $HgSO_4^0$, HgS_2^{2-}, $Hg(HS)_2^0$, $HgNH_3^{2+}$, $Hg(NH_3)_2^{2+}$, $Hg(NH_3)_3^{2+}$, $Hg(NH_3)_4^{2+}$, $HgNO_3^+$, $Hg(NO_3)_2^0$
Thallium (Tl)	Tl^+, $Tl(OH)_3^0$, $TlOH^0$, Tl^{3+}, $TlOH^{2+}$, $Tl(OH)_2^+$, $Tl(OH)_4^-$, TlF^0, $TlCl^0$, $TlCl_2^-$, $TlCl^{2+}$, $TlCl_2^+$, $TlCl_3^0$, $TlCl_4^-$, $TlOHCl^+$, $TlBr^0$, $TlBr_2^-$, $TlBrCl^-$, $TlBr^{2+}$, $TlBr_2^+$, $TlBr_3^0$, $TlBr_4^-$, TlI^0, TlI_2^-, $TlIBr^-$, TlI_4^-, $TlSO_4^-$, $TlHS^0$, Tl_2HS^+, $Tl_2OH(HS)_2^-$, $Tl_2(OH)_2(HS)_2^-$, $TlNO_3^0$, $TlNO_2^0$, $TlNO_3^{2+}$
Lead (Pb)	Pb^{2+}, $PbOH^+$, $Pb(OH)_2^0$, $Pb(OH)_3^-$, Pb_2OH^{3+}, $Pb_3(OH)_4^{2+}$, $Pb(OH)_4^{2-}$, PbF^+, PbF_2^0, PbF_3^-, PbF_4^{2-}, $PbCl^+$, $PbCl_2^0$, $PbCl_3^-$, $PbCl_4^{2-}$, $PbBr^+$, $PbBr_2^0$, PbI^+, PbI_2^0, $PbSO_4^0$, $Pb(SO_4)_2^{2-}$, $Pb(HS)_2^0$, $Pb(HS)_3^-$, $PbNO_3^+$, $Pb(CO_3)_2^{2-}$, $PbCO_3^0$, $PbHCO_3^+$
Thorium (Th)	Th^{4+}, ThF^{3+}, ThF_2^{2+}, ThF_3^+, ThF_4^0, $Th(OH)_2^{2+}$, $Th(OH)^{3+}$, $Th(OH)_4^0$, $Th_2(OH)_2^{6+}$, $Th_4(OH)_8^{8+}$, $Th_6(OH)_{15}^{9+}$, $ThOH^{3+}$, $ThCl^{3+}$, $ThCl_2^{2+}$, $ThCl_3^+$, $ThCl_4^0$, $Th(H_2PO_4)_2^{2+}$, $Th(HPO_4)_2^0$, $Th(HPO_4)_3^{2-}$, $ThH_2PO_4^{3+}$, $ThH_3PO_4^{4+}$, $ThHPO_4^{2+}$, $Th(SO_4)_2^0$, $Th(SO_4)_3^{2-}$, $Th(SO_4)_4^{4-}$, $ThSO_4^{2+}$
Radium (Ra)	Ra^{2+}, $RaOH^+$, $RaCl^+$, $RaSO_4^0$, $RaCO_3^0$, $RaHCO_3^+$
Uranium (U)	U^{4+}, UOH^{3+}, $U(OH)_2^{2+}$, $U(OH)_3^+$, $U(OH)_4^0$, $U(OH)_5^-$, $U_6(OH)_{15}^{9+}$, UO_2OH^+, $(UO_2)_2(OH)_2^{2+}$, $(UO_2)_3(OH)_5^+$, UO_2^{2+}, UF^{3+}, UF_2^{2+}, UF_3^+,

UF_4^0, UF_5^-, UF_6^{2-}, UO_2F^+, $UO_2F_2^0$, $UO_2F_3^-$, $UO_2F_4^{2-}$, UCl^{3+}, UO_2Cl^+, USO_4^{2+}, $U(SO_4)_2^0$, $UO_2SO_4^0$, $UO_2(SO_4)_2^{2-}$, $UHPO_4^{2+}$, $U(HPO_4)_2^0$, $U(HPO_4)_3^{2-}$, $U(HPO_4)_4^{4-}$, $UO_2HPO_4^0$, $UO_2(HPO_4)_2^{2-}$, $UO_2H_2PO_4^+$, $UO_2(H_2PO_4)_2^0$, $UO_2(H_2PO_4)_3^-$, $UO_2H_2AsO_4^+$, $UO_2HAsO_4^0$, $UO_2(H_2AsO_4)_2^0$, $UO_2CO_3^0$, $UO_2(CO_3)_2^{2-}$, $UO_2(CO_3)_3^{4-}$, $Ca_2(UO_2)(CO_3)_3^0$, $UO_2H_3SiO_4^+$

Besides inorganic species, numerous organic (Table 2) and organisms (Table 3) are encountered in water that are of great importance for water quality.

Table 2   Selected organic substances (plus-sign in brackets means that geogenic formation in traces is possible, only the typical concentration range is indicated)

Substance	geogenic	anthropogenic	typical range of concentration
Humic matter	+	-	mg/L
aliphatic carbons: oil, fuel	+	+	mg/L
Phenols	+	+	mg/L
BTEX (benzene, toluene, ethylbenzene, xylene)	(+)	+	µg/L
PAHs (polycyclic aromatic hydrocarbons)	(+)	+	µg/L
PCBs (polychlorinated biphenyls)	-	+	µg/L
CFC´s (Chlorofluorocarbons)	-	+	ng/L
Dioxins, furans	(+)	+	pg/L
pesticides	(+)	+	ng/L
hormones	(+)	+	pg/L
pharmaceuticals	-	+	pg/L

Table 3   Organisms in groundwater

	size
Virus	5 - 300 nm
Prokaryotes: Bacteria & Archaea (methaneogenous, extreme halophiles, extreme thermophiles)	100 - 15.000 nm
Eukaryotes: Protozoa (Foraminifera, Radiolaria, Dinoflagellata) Yeast (anaerob) Fungi (aerob)	> 3 µm ~20 µm
Fish (Brotulidae, Amblyopsidae, Astyanax Jordani, Caecobarbus Geertsi) in Karst aquifers	mm… cm dm… m

Interactions of different species within the aqueous phase (chapter 1.1.5), with gases (chapter 1.1.3), and solid phases (minerals) (chapter 1.1.4.) as well as transport (chapter 1.3) and decay processes (biological decomposition, radioactive decay) are fundamental in determining the hydrogeochemical composition of ground and surface water.

Hydrogeochemical reactions involving only a single phase are called homogeneous, whereas heterogeneous reactions occur between two or more phases such as gas and water, water and solids, or gas and solids. In contrast to open systems, closed systems enable only exchange of energy, not constituents with the surrounding environment.

Chemical reactions can be described by thermodynamics (chapter 1.1.2) and kinetics (chapter 1.2). Reactions expressed by the mass-action law (chapter 1.1.2.1), are thermodynamically reversible and independent of time. In contrast, kinetic processes are time-dependent reactions. Models that take into account kinetics can describe irreversible reactions such as decay processes that require finite amounts of time.

1.1.2 Thermodynamic fundamentals

1.1.2.1 Mass-action law

In principle, any chemical equilibrium can be described by the mass-action law.

$$aA + bB \leftrightarrow cC + dD \hspace{5cm} \text{Eq. (1.)}$$

$$K = \frac{\{C\}^c \cdot \{D\}^d}{\{A\}^a \cdot \{B\}^b} \hspace{5cm} \text{Eq. (2.)}$$

With a, b, c, d = number of moles of the reactants A, B, and the end products C, D, respectively for the given reaction, (1);

K = thermodynamic equilibrium or dissociation constant (general term)

In particular, the term K is defined in relation to the following types of reactions:
- Dissolution/precipitation (chapter 1.1.4.1)
    K_S = solubility-product
- Sorption (chapter 1.1.4.2)
    K_d = distribution coefficient
    K_x = selectivity coefficient
- Complex formation/dissolution of complexes (chapter 1.1.5.1)
    K = complexation constant, stability constant
- Redox reaction (chapter 1.1.5.2)
    K = stability constant

If one reverses reactants and products in a reaction equation, the solubility constant K' equals 1/K. Hence, it is important when reporting an equilibrium constant to always refer to the corresponding reaction equation. Furthermore, one must distinguish between conditional constants, which are valid only for certain conditions of temperature and ionic strength, and generally applicable constants, at standard state conditions (i.e. T=25°C and ionic strength I=0, infinitely diluted solution). The latter might be calculated from the former. Standard temperature conditions can be calculated using the van't Hoff equation (Eq. 3):

$$\log(K_r) = \log(K_0) + \frac{H^0_r}{2.303 \cdot R} \cdot \frac{T_K - T_{K_0}}{T_K \cdot T_{K_0}}$$

Eq. (3.)

with K_r = equilibrium constant at measured temperature

          K_0 = equilibrium constant at standard temperature

          T_K = measured temperature in degrees Kelvin

          T_{K0} = temperature in Kelvin, at which the standard enthalpy H^0_r was determined

          R   = ideal gas constant (8.315 J/K mol)

Transformation to standard pressure is done using the following equation (Eq. 4):

$$\ln K(P) = \ln K(S) - \frac{\Delta V(T)}{T \cdot R \cdot \beta} \cdot \ln \frac{\sigma(P)}{\sigma(S)}$$

Eq. (4.)

with $K(P)$ = equilibrium constant at pressure P

          $K(S)$ = equilibrium constant at saturated vapor pressure

          $\Delta V(T)$ = volume change of the dissociation reaction at temperature T and saturated vapor pressure S

          β = isothermal compressibility coefficient of water at T and P

          $\sigma(P)$ = density of water at pressure P

          $\sigma(S)$ = density of water at saturated vapor pressure

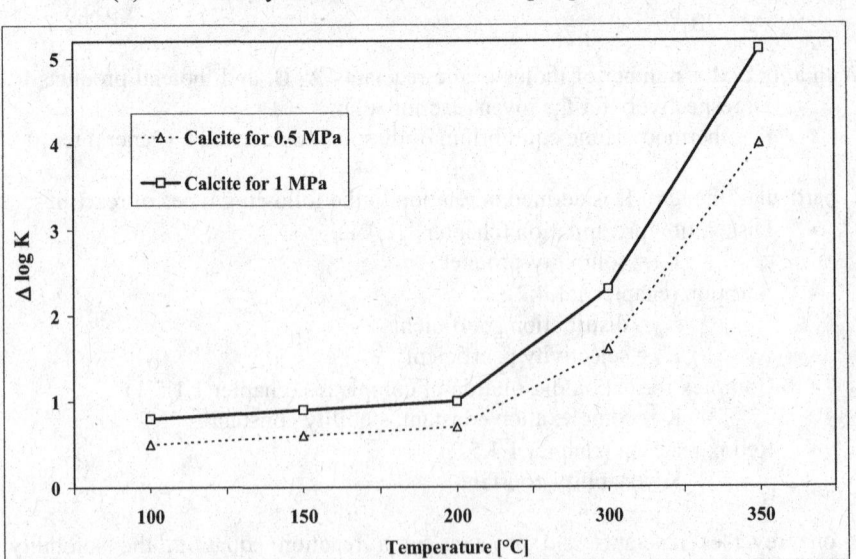

Fig. 1   Influence of pressure and temperature on the solubility of calcite (after Kharaka et al. 1988)

Fig. 1 shows calcite dissolution dependent on different pressure and temperature conditions.

If a process consists of a series of subsequent reactions, as for instance the dissociation of H_2CO_3 to HCO_3^- and to CO_3^{2-}, stability (dissociation) constants are numbered consecutively (e.g. K_1 and K_2).

1.1.2.2 Gibbs free energy

A system at constant temperature and pressure is in disequilibrium until all of its Gibbs free energy, G, is consumed. At equilibrium, Gibbs free energy is zero.

The Gibbs free energy is a measure of the probability that a reaction occurs. It is defined by the enthalpy, H, and the entropy, S^0 (Eq. 5). The enthalpy can be described as the thermodynamic potential, which ensues $H = U + p*V$, where U is the internal energy, p is the pressure, and V is the volume. The entropy, according to classical definitions, is a measure of molecular order of a thermodynamic system and the irreversibility of a process, respectively.

$$G = H - S^0 \cdot T$$
Eq. (5.)

with T = temperature in Kelvin

A positive value for G means that additional energy is required for the reaction to happen, and a negative value means that the reaction occurs spontaneously, thereby releasing energy.

The change in free energy of a reaction is directly related to the change in energy of the activities of all reactants and products under standard conditions.

$$G = G^0 + R \cdot T \cdot \ln \frac{\{C\}^c \cdot \{D\}^d}{\{A\}^a \cdot \{B\}^b}$$
Eq. (6.)

with R = ideal gas constant
G^0 = standard Gibbs free energy at 25°C and 100 kPa

G^0 equals G, when all reactants have the same activity. Then, the term within the logarithm in Eq. 6 is 1 and the logarithm becomes zero.

The equation can further be simplified in equilibrium, when G equals zero, to:

$$G^0 = -R \cdot T \cdot \ln K$$
Eq. (7.)

Accordingly G enables a prediction on the equilibrium of the reaction aA + bB $\leftrightarrow$ cC + dD. With G <0, reactants will be consumed while the concentration of products increases, for G>0 product are consumed and reactants formed.

1.1.2.3 Gibbs phase rule

The Gibbs phase rule gives the number of degrees of freedom that result from the number of components and phases, coexisting in a system.

$$F = C - P + 2 \hspace{4cm} \text{Eq. (8.)}$$

with   F = number of degrees of freedom
         C = number of components
         P = number of phases

The number 2 in the Eq. 8 arises from the two independent variables, pressure and temperature. Phases are limited, physically and chemically homogeneous, mechanically separable parts of a system. Components are defined as simple chemical entities or units that comprise the composition of a phase.

In a system, where the number of phases and the number of components are equal, there are two degrees of freedom, i.e. that two variables can be varied independently (e.g. temperature and pressure). If the number of the degrees of freedom is zero, temperature and pressure are constant and the system is invariant.

In a three-phase system including a solid and a liquid as well as a gas, the Gibbs phase rule is modified to:

$$F = C' - N - P + 2 \hspace{3.5cm} \text{Eq. (9.)}$$

with   F = number of the degrees of freedom
         C' = number of different chemical species
         N = number of possible equilibrium reactions (species, charge balance, stoichiometric relations)
         P = number of phases

1.1.2.4 Activity

Using the mass-action law, quantities of substances are represented as activities, a_i, and not as concentrations, c_i, with respect to a species, i.

$$a_i = f_i \cdot c_i \hspace{4cm} \text{Eq.(10.)}$$

In Eq. 10, the activity coefficient, f_i, is an ion-specific correction factor that describes interactions among charged ions. Since the activity coefficient is a non-linear function of ionic strength, the activity is a non-linear function of the concentration, too.

The activity decreases with increasing ionic strength up to 0.1 mol/kg and is always lower than the concentration, because charged ions reduce each other's activities in solution through interactions. Thus, the activity coefficient is less than 1 (Fig. 2). It is obvious from Fig. 2 that when increasing the ion concentrations the decrease in activity is the more significant the higher the valence of the ions. In the ideal case of an infinitely dilute solution, where ion-interactions are close to zero, the activity coefficient is 1 and the activity equals the concentration.

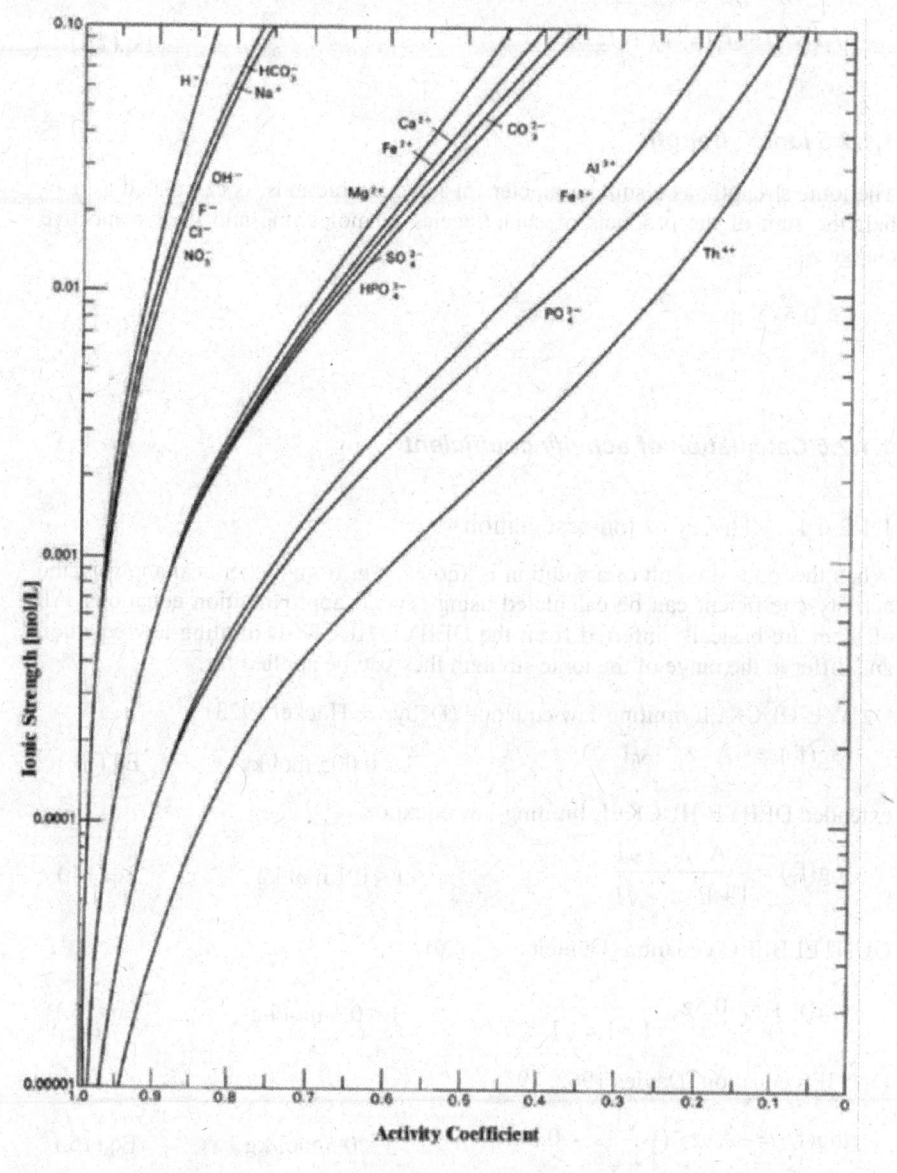

Fig. 2   Relation between ionic strength and activity coefficient in a range up to 0.1 mol/L (after Hem 1985)

Experimentally, only mean activity coefficients can be determined for salts, not activity coefficients for single ions. According to the MacInnes Convention, however, it can be assumed that cations and anions formed upon dissolution of one compound have the same activity coefficients because of their same charge, similar size, electron configuration and mobility, e.g.:

$$f_i(K^+) = f_i(Cl^-) = f_\pm(KCl)$$ Eq.(11.)

1.1.2.5 Ionic strength

The ionic strength, as a sum parameter for ionic interactions, is calculated as one-half the sum of the products of each species in moles, m_i, and their respective charge z_i.

$$I = 0.5 \cdot \sum_i m_i \cdot z_i^2$$ Eq.(12.)

1.1.2.6 Calculation of activity coefficient

1.1.2.6.1. Theory of ion-association

When the ionic strength of a solution is known, e.g. from a chemical analysis, the activity coefficient can be calculated using several approximation equations. All of them are basically inferred from the DEBYE-HÜCKEL limiting-law equation and differ in the range of the ionic strength they can be applied for.

DEBYE-HÜCKEL limiting-law equation (Debye & Hückel 1923)

$$\log(f_i) = -A \cdot z_i^2 \cdot \sqrt{I}$$ $I < 0.005$ mol/kg   Eq.(13.)

extended DEBYE-HÜCKEL limiting-law equation

$$\log(f_i) = \frac{-A \cdot z_i^2 \cdot \sqrt{I}}{1 + B \cdot a_i \cdot \sqrt{I}}$$ $I < 0.1$ mol/kg   Eq.(14.)

GÜNTELBERG equation (Güntelberg 1926)

$$\log(f_i) = -0.5 z_i^2 \frac{\sqrt{I}}{1 + 1.4\sqrt{I}}$$ $I < 0.1$ mol/kg   Eq.(15.)

DAVIES equation (Davies 1962, 1938)

$$\log(f_i) = -A \cdot z_i^2 (\frac{\sqrt{I}}{1 + \sqrt{I}} - 0.3 \cdot I)$$ $I < 0.5$ mol/kg   Eq.(16.)

"WATEQ" DEBYE-HÜCKEL equation (Hückel 1925)

$$\log(f_i) = \frac{-A \cdot z_i^2 \cdot \sqrt{I}}{1 + B \cdot a_i \cdot \sqrt{I}} + b_i \cdot I$$ $I < 1$ mol/kg   Eq.(17.)

with   f = activity coefficient
          z = valence
          I = ionic strength

a_i, b_i = ion-specific parameters (dependent on the ion radius) (selected values see Table 4, complete overview in van Gaans (1989) and Kharaka et al. (1988))

A,B temperature-dependent parameters, calculated from the following empirical equations (Eq. 18 to Eq. 21)

$$A = \frac{1.82483 \cdot 10^6 \cdot \sqrt{d}}{(\varepsilon \cdot T_K)^{3/2}} \qquad \text{Eq.(18.)}$$

$$B = \frac{50.2916 \cdot \sqrt{d}}{(\varepsilon \cdot T_K)^{1/2}} \qquad \text{Eq.(19.)}$$

$$d = 1 - \frac{(T_c - 3.9863)^2 \cdot (T_c + 288.9414)}{508929.2 \cdot (T_c + 68.12963)} + 0.011445 \cdot e^{-374.3/T_c} \qquad \text{Eq.(20.)}$$

$$\varepsilon = 2727.586 + 0.6224107 \cdot T_K - 466.9151 \cdot \ln(T_K) - \frac{52000.87}{T_K} \qquad \text{Eq.(21.)}$$

with   d = density (after Gildseth et al. 1972 for 0-100°)
        ε = dielectric constant (after Nordstrom et al. 1990 for 0-100°C)
        T_C = temperature in °Celsius
        T_K = temperature in Kelvin

For temperatures of about 25°C and water with a density of 1 g/cm³, A is 0.51 and B is 0.33. In some textbooks, B is charted as $0.33 \cdot 10^8$. For the use of the latter, a_i must be in cm, otherwise in Å ($=10^{-8}$ cm).

Table 4   Ion-specific parameters a_i and b_i (after Parkhurst et al. 1980 and (*) Truesdell a. Jones 1974)

Ion	a_i [Å]	b_i [Å]	Ion	a_i [Å]	b_i [Å]
H^+	4.78	0.24	Mn^{2+}	7.04	0.22
Li^+	4.76	0.20	Fe^{2+}	5.08	0.16
Na^+ (*)	4.0	0.075	Co^{2+}	6.17	0.22
Na^+	4.32	0.06	Ni^{2+}	5.51	0.22
K^+ (*)	3.5	0.015	Zn^{2+}	4.87	0.24
K^+	3.71	0.01	Cd^{2+}	5.80	0.10
Cs^+	1.81	0.01	Pb^{2+}	4.80	0.01
Mg^{2+} (*)	5.5	0.20	OH^-	10.65	0.21
Mg^{2+}	5,46	0.22	F^-	3.46	0.08
Ca^{2+} (*)	5.0	0.165	Cl^-	3.71	0.01
Ca^{2+}	4.86	0.15	ClO_4^-	5.30	0.08
Sr^{2+}	5.48	0.11	HCO_3^-, CO_3^{2-} (*)	5.40	0
Ba^{2+}	4.55	0.09	SO_4^{2-} (*)	5.0	-0.04
Al^{3+}	6.65	0.19	SO_4^{2-}	5.31	-0.07

The theory of ion association, using only ionic strength-dependent activity coefficients, is valid to about 1 molal. Above that concentration, it has not been tested. Some authors even believe the upper limit should be at 0.7 mol/kg (which equals sea water). Fig. 3 shows that already at an ionic strength of > 0.3 mol/kg (H^+) the activity coefficient does not decrease any further but starts to increase, and eventually attains values of more than 1. The second term in the DAVIES and extended DEBYE-HÜCKEL limiting-law equations forces the activity coefficient to increase at high ionic strength. This is owed to the fact, that ion-interactions are not only based on Coulomb forces any more, ion sizes change with the ionic strength, and ions with the same charge interact. Moreover, with the increase in the ionic strength, a larger fraction of water molecules is bound to ion hydration sleeves decreasing the concentration of free water molecules and correspondingly increasing the activity relative to 1kg of free water molecules.

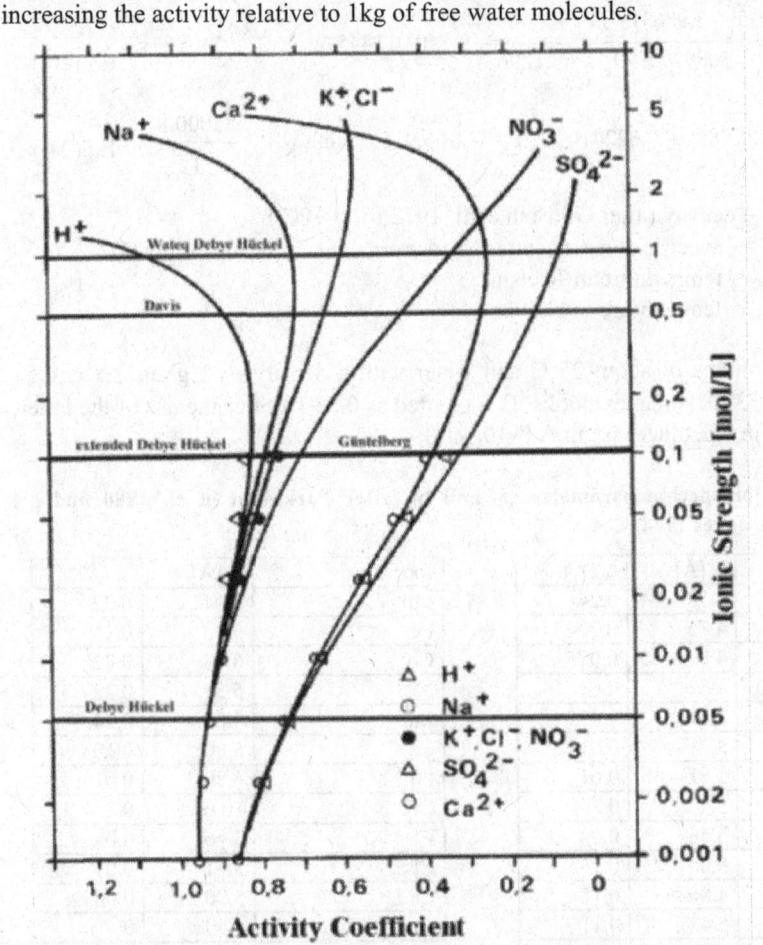

Fig. 3   **Relation of ionic strength and activity coefficient in higher concentrated solutions, (up to I = 10 mol/kg), valid ranges for the different electrolyte theories are indicated as lines (modified after Garrels and Christ 1965)**

1.1.2.6.2. Theory of ion-interaction

For higher ionic strengths, e.g. highly saline waters, the PITZER equation can be used (Pitzer 1973). This semi-empirical model is based also on the DEBYE-HÜCKEL limiting-law equation, but additionally integrates "virial" equations (vires = Latin for forces), that describe ion-interactions (intermolecular forces). Compared to the ion-association theory, parameters for PITZER equation are often lacking for more complex species. Furthermore, a set of equilibrium constants (albeit minimal) for complexation reactions is still required.

In the following a simple version of PITZER equations is briefly described. For complete calculations and description of specific parameters and equations the reader is referred to the original literature (Pitzer 1973, Pitzer 1981, Whitfield 1975, Whitfield 1979, Silvester and Pitzer 1978, Harvie and Weare 1980, Gueddari et al. 1983, Pitzer 1991).

The calculation of the activity coefficient is done separately for positively (index i) and negatively (index j) charged species applying Eq. 22. In the following example the calculation of the activity coefficients is shown for cations; the calculation for anions works correspondingly.

$$\ln f_M = z_M^2 \cdot F + S1 + S2 + S3 + \left| z_M \right| \cdot S4 \qquad\qquad \text{Eq.(22.)}$$

with   M = cation

z_M = valence state of cation M

F, S1-S4 = sums, calculated using Eqs. 23-30

$$S1 = \sum_{j=1}^{a} m_j (2 \cdot B_{Mj} + z \cdot C_{Mj}) \qquad\qquad \text{Eq.(23.)}$$

$$S2 = \sum_{i=1}^{c} m_i (2 \cdot \phi_{Mj} + \sum_{j=1}^{a} m_j \cdot P_{Mij}) \qquad\qquad \text{Eq.(24.)}$$

$$S3 = \sum_{j=1}^{a-1} \sum_{k=j+1}^{a} m_j^2 \cdot P_{Mjk} \qquad\qquad \text{Eq.(25.)}$$

$$S4 = \sum_{i=1}^{c} \sum_{j=1}^{a} m_i \cdot m_j \cdot c_{ij} \qquad\qquad \text{Eq.(26.)}$$

with   B, C, Φ, P = species-specific parameters

m = molarities [mol/L]

k = index

c = number of cations

a = number of anions

$$F = -\frac{2.303 \cdot A}{3.0} (\frac{\sqrt{I}}{1+1.2 \cdot \sqrt{I}} + \frac{2}{1.2} \cdot \ln(1+1.2 \cdot \sqrt{I})) + S5 + S6 + S7 \qquad \text{Eq.(27.)}$$

$$S5 = \sum_{i=1}^{c} \sum_{j=1}^{a} m_i \cdot m_j \cdot B'_{ij} \qquad\qquad Eq.(28.)$$

$$S6 = \sum_{i=1}^{c-1} \sum_{k=i+1}^{c} m_i^2 \cdot \phi'_{ik} \qquad\qquad Eq.(29.)$$

$$S7 = \sum_{j=1}^{a-1} \sum_{l=j+1}^{a} m_j^2 \cdot \phi'_{jl} \qquad\qquad Eq.(30.)$$

with A = DEBYE-HÜCKEL constant (Eq. 18)
        B', Φ' = Virial coefficients, modified with regard to the ionic strength
        k, l = indices

1.1.2.7 Comparison ion-association versus ion-interaction theory

Fig. 4 to Fig. 8 show the substantial differences in activity coefficients for calcium, chloride, sulfate, sodium, and hydrogen calculated based on different ion-association and interaction theories. Activity coefficients were calculated according to Eq. 13 to Eq. 17 for the corresponding ion-association theories. With the program PHRQPITZ activity coefficients were calculated based on the PITZER ion-interaction model. The substantial differences clearly show the limited validity of the different theories.

In particular, the strongly diverging graph of the simple DEBYE-HÜCKEL limiting-law equation from the PITZER curve in the range exceeding 0.005 mol/kg (validity limit) is obvious. In contrast, the conformity of WATEQ-DEBYE-HÜCKEL and PITZER concerning the divalent calcium and sulfate ions is surprisingly good. Also for chloride the WATEQ-DEBYE-HÜCKEL and PITZER equation show a good agreement up to I = 3 mol/kg. On contrary, the activity coefficients for sodium and hydrogen ions show substantial discrepancies. There, the validity range of 1 mol/kg for the WATEQ-DEBYE-HÜCKEL limiting-law equation must be restricted, since significant differences already occur at ionic strengths as low as 0.1 mol/kg (one order of magnitude below the cited limit) in comparison to the PITZER equation. These examples demonstrate the flaws of the ion-association theory, which are especially severe for the mono-valent ions.

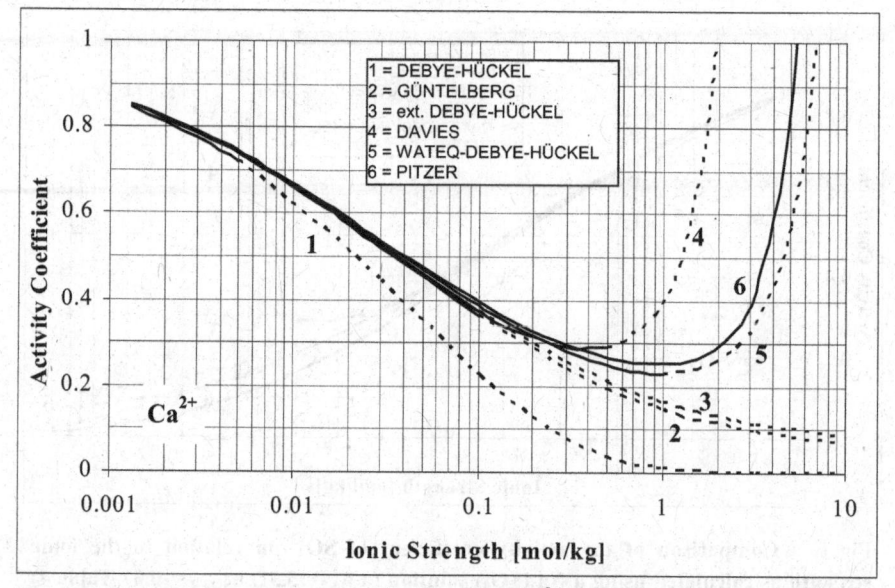

Fig. 4   Comparison of the activity coefficient of Ca^{2+} in relation to the ionic strength as calculated using a $CaCl_2$ solution ($a_{Ca} = 4.86$, $b_{Ca} = 0.15$ Table 4) and different theories of ion-association and the PITZER equation, dashed lines signify calculated coefficients outside the validity range of the corresponding ion-association theory.

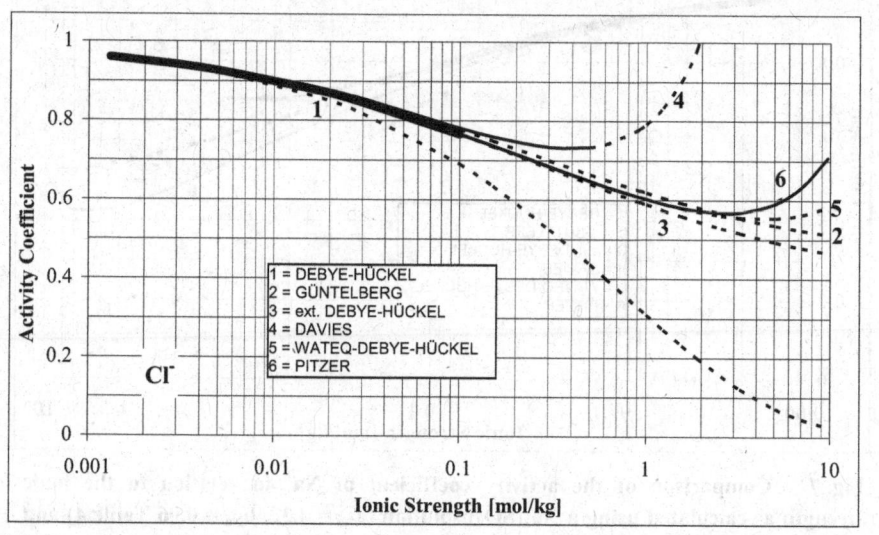

Fig. 5   Comparison of the activity coefficient of Cl^- in relation to the ionic strength as calculated using a $CaCl_2$ solution ($a_{Cl} = 3.71$, $b_{Cl} = 0.01$ Table 4) and different theories of ion-association and the PITZER equation, dashed lines signify calculated coefficients outside the validity range of the corresponding ion-association theory.

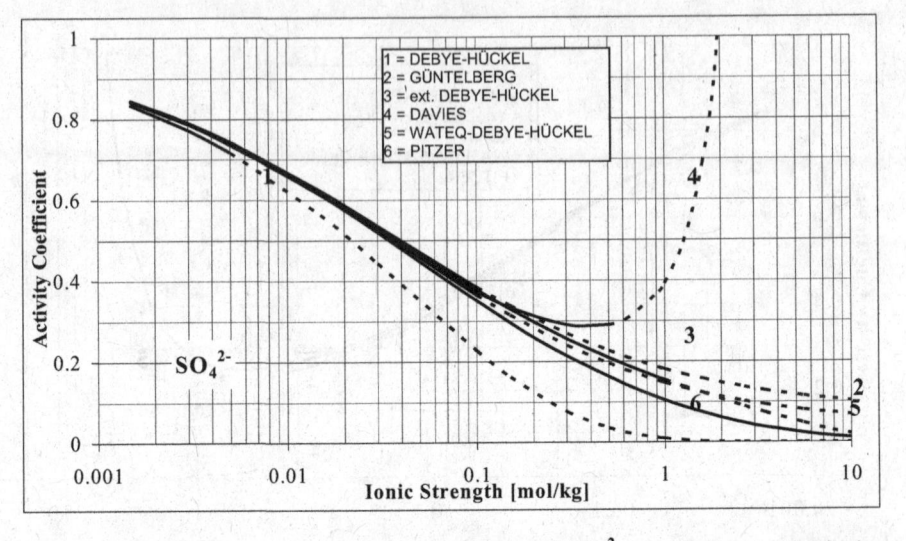

Fig. 6 Comparison of the activity coefficient of SO_4^{2-} in relation to the ionic strength as calculated using a $Na_2(SO_4)$ solution ($a_{SO4-2}= 5.31$, $b_{SO4-2}= -0.07$ Table 4) and different theories of ion-association and the PITZER equation, dashed lines signify calculated coefficients outside the validity range of the corresponding ion-association theory.

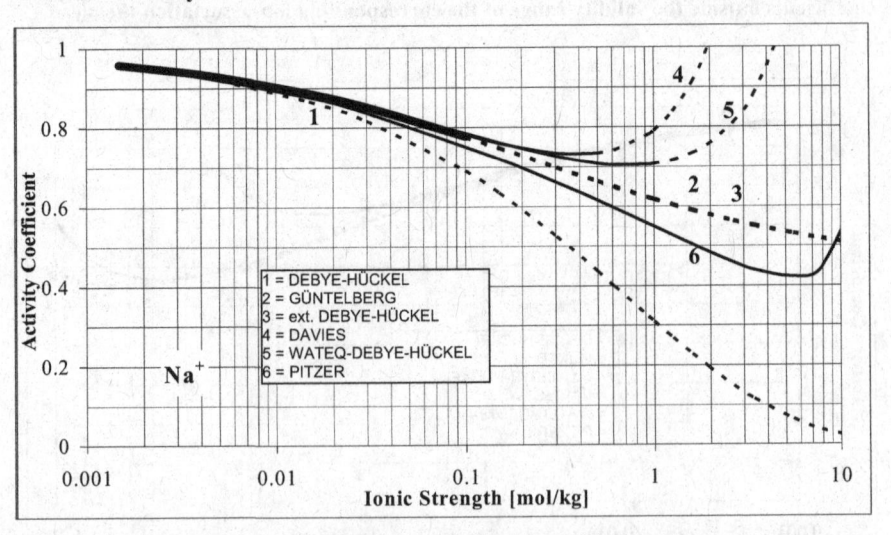

Fig. 7 Comparison of the activity coefficient of Na^+ in relation to the ionic strength as calculated using a $Na_2(SO_4)$ solution ($a_{Na} = 4.32$, $b_{Na} = 0.06$ Table 4) and different theories of ion-association and the PITZER equation, dashed lines signify calculated coefficients outside the validity range of the corresponding ion-association theory.

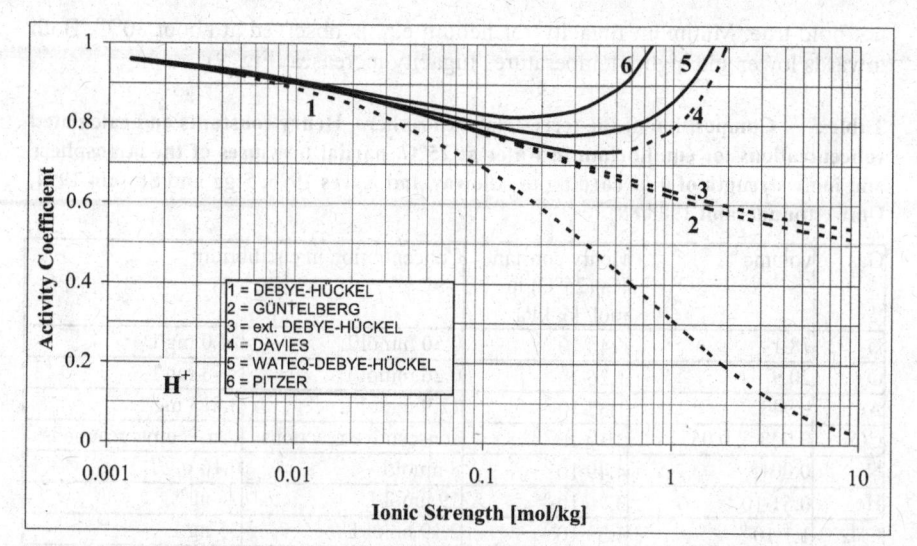

Fig. 8 **Comparison of the activity coefficient of H^+ in relation to the ionic strength as calculated from the changing pH of a $CaCl_2$ solution ($a_H = 4.78$, $b_H = 0.24$ Table 4) using different theories of ion-association and the PITZER equation, dashed lines signify calculated coefficients outside the validity range of the corresponding ion-association theory.**

1.1.3 Interactions at the liquid-gaseous phase boundary

1.1.3.1 Henry Law

Using the linear Henry's law, the amount of gas dissolved in water can be calculated for a known temperature and partial pressure.

$$a_i = K_{Hi} \cdot p_i \qquad\qquad\qquad\qquad \text{Eq.(31.)}$$

a_i = fugacity (activity) of the gas [mol/kg]
K_{Hi} = Henry constant of the gas i
p_i = partial pressure of the gas i[kPa]

Table 5 shows the Henry constants and the inferred amount of gas dissolved in water for different gases of the atmosphere. The partial pressures of N_2 and O_2 in the atmosphere at 25°C and 10^5Pa (1 bar), for example, are 78 kPa and 21 kPa, respectively. These pressures correspond to concentrations of 14.00 mg/L for N_2 and 8.43 mg/L for O_2.

Gas solubility generally decreases with increasing temperature. For oxygen and nitrogen, e.g. solubility at 100°C is only 45 and 54%, respectively, of that at 10°C. However, in particular for light gases like H and He this inverse correlation does not hold true. Minimum fugacity for helium e.g. is observed at about 30°C. Both towards lower and higher temperatures fugacity increases (Fig. 9).

Table 5   Composition of the terrestrial atmosphere, Henry constants and calculated concentrations for equilibrium in water at 25°C, partial pressures of the atmosphere and ionic strength of 0 (according to Alloway and Ayres 1996, Sigg and Stumm 1994, Umweltbundesamt 1988/89).

Gas	volume %	Henry constant K_H (25°C) in mol/ kg·kPa	Concentration in equilibrium	
N_2	78.1	$6.40·10^{-6}$	0.50 mmol/L	14.0 mg/L
O_2	20.9	$1.26·10^{-5}$	0.26 mmol/L	8.43 mg/L
Ar	0.943	$1.37·10^{-5}$	12.9 mmol/L	0.515 mg/L
CO_2	0.028 ... 0.037	$3.39·10^{-4}$	consecutive reactions	consecutive reactions
Ne	0.0018	$4.49·10^{-6}$	8 nmol/L	0.16 mg/L
He	$0.51·10^{-3}$	$3.76·10^{-6}$	19 nmol/L	76 ng/L
CH_4	$1.7·10^{-6}$	$1.29·10^{-5}$	2.19 nmol/L	35 ng/L
N_2O	$0.304·10^{-6}$	$2.57·10^{-4}$	0.078 nmol/L	3.4 ng/L
NO	---	$1.9·10^{-5}$	consecutive reactions	consecutive reactions
NO_2	$10 .... 22·10^{-9}$	$1.0·10^{-4}$	consecutive reactions	consecutive reactions
NH_3	$0.2-2·10^{-9}$	0.57	consecutive reactions	consecutive reactions
SO_2	$10·10^{-9} ... 19·10^{-9}$	0.0125	consecutive reactions	consecutive reactions
O_3	$10·10^{-9} ... 100·10^{-9}$	$9.4·10^{-5}$	0.094 ... 0.94 nmol/L	4.5 ... 45 ng/L

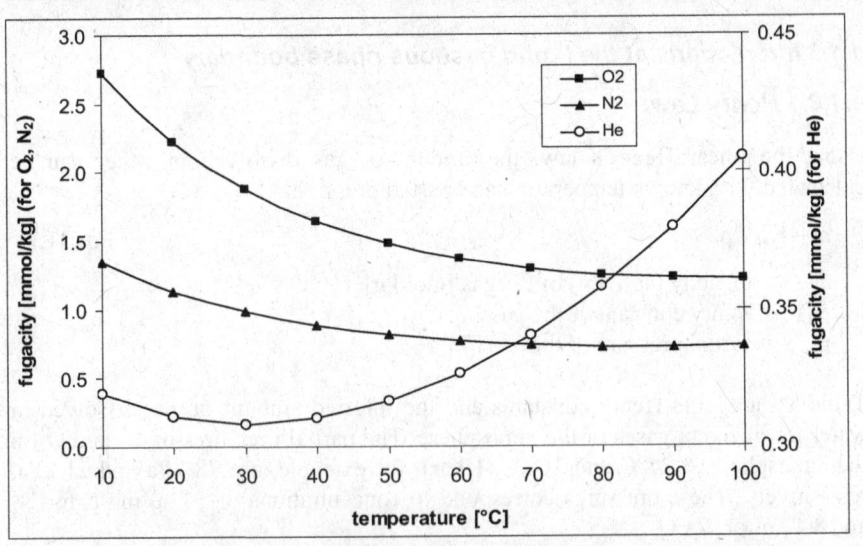

Fig. 9   Fugacity of different gases with increasing temperature. While the fugacity of oxygen and nitrogen decreases with increasing temperature, helium shows a minimum fugacity at 30°C, increasing both towards lower and higher temperatures (modeled with PHREEQC, based on the LLNL database)

Thus, Henry's law is only directly applicable for gases, which do not undergo spontaneous further reactions, as for example nitrogen, oxygen, or argon. For gases that react with water, Henry's law can only be applied if subsequent reactions are taken into account. One such example where solubility is much higher than calculated by Henry's Law is carbon dioxide. Even though, dependent on pH, only about 1% of CO_2 spontaneously dissolve in water to form HCO_3^- and CO_3^{2-}, subsequent complexation reactions of HCO_3^- and CO_3^{2-} with mono- or divalent metals strongly increase the aqueous solubility of CO_2. Consumption of protons during mineral dissolution (e.g. calcite) further increases dissolution of CO_2.

1.1.4 Interactions at the liquid-solid phase boundary

1.1.4.1 Dissolution and precipitation

Dissolution and precipitation can be described by the mass-action law as reversible, heterogeneous reactions. In general, the solubility of a mineral is defined as the mass of a mineral, which can be dissolved within a standard volume of the solvent.

1.1.4.1.1. Solubility-product

The dissolution of a mineral AB into the components A and B occurs according to the mass-action law as follows:

$$AB \leftrightarrow A + B \qquad\qquad \text{Eq.(32.)}$$

$$K_{sp} = \frac{\{A\} \cdot \{B\}}{\{AB\}} \qquad\qquad \text{Eq.(33.)}$$

Since the activity of the solid phase AB is assumed to be constant at 1, the equilibrium constant of the mass-action law results in the following solubility-product constant (K_{sp}) or ion-activity product (IAP):

$$K_{sp} = IAP = \{A\} \cdot \{B\} \qquad\qquad \text{Eq.(34.)}$$

It is important to note that analytically determined concentrations for A and B must be transformed into activities and that only the activities of the free ions are considered, not those of the complexes.

The solubility-product depends on the mineral, the solvent, the pressure or the partial pressure of certain gases, the temperature, pH, E_H and on the ions previously dissolved in the water and to what extent these have already formed complexes. While partial pressure, pH, E_H, and complex stability are considered in the mass-action law, temperature and pressure have to be taken into account by additional factors.

Dependency of K_{SP} on the pressure
Up to a pressure equivalent to 500 m water depth (5 MPa) total pressure changes have almost no influence on the solubility-product. There is, however, a strong dependency on the partial pressure of particular gases.

Dependency of K_{SP} on the partial pressure
An increased carbon dioxide partial pressure is the main reason for the observed increased rate of dissolution and precipitation in the upper layer of soils. In the growth season the carbon dioxide partial pressure can be about 10 to 100 times higher than in the atmosphere because of the biological and microbiological activity. While the average carbon dioxide partial pressure under humid climate conditions in summer is at 3 to 5 kPa (3-5 vol%), it amounts to up to 30 vol% in tropical climates and to up to 60 vol% in heaps or organically contaminated areas. Since the increased partial pressure of CO_2 is accompanied by a higher proton activity, minerals with increased solubility under acidic conditions are preferably dissolved.

Dependency of K_{SP} on the temperature
In contrast to the partial pressure, temperature rise does not generally contribute to an increase in solubility. According to the principle of the smallest constraint (Le Chatelier), only endothermic dissolutions, i.e. reactions, which need additional heat input, are favored (e.g. dissolution of silicates, aluminosilicates, oxides, etc.). The dissolutions of carbonates and sulfates which are exothermic reactions decrease with increasing temperature.

Dependency of K_{SP} on the pH value
Few ions are equally soluble over the whole range of pH conditions in natural groundwaters. Especially the dissolution of metals is strongly pH-dependent. While precipitating as hydroxides, oxides, and salts under basic conditions, they dissolve under acid conditions and are highly mobile as free cations. Aluminum is soluble under acid as well as under basic conditions. It precipitates as hydroxide or clay mineral in the pH range of 5 to 8.

Dependency of K_{SP} on the E_H value
For redox-sensitive elements, i.e., elements that occur in different oxidation states depending on the prevailing oxidizing or reducing conditions, the solubility does not only depend on the pH only but also on the redox chemistry. For example, uranium in its reduced form as U(4) is almost insoluble at moderate pH values, while U(6) is readily dissolved. For iron (at pH > 3) it is vice versa: Substantial amounts can only be dissolved in its reduced form as Fe(2), while Fe(3) rapidly precipitates as ironhydroxides even at low concentrations in solution.

Table 6 Periodic system depicting the relative enrichment (ratio > 1) of elements in sea water as compared to river water; elements enriched in sea water (mobile elements) are shaded (after Faure 1991, Merkel and Sperling 1996, 1998)

1	2	3	4	5	6	7	8	9	10	11	12	13	14	15	16	17	18
H —																	He
Li 56.7	Be 0.02											B 450	C —	N —	O —	F 1300	Ne
Na 1714	Mg 315											Al 0.016	Si 0.43	P 3.6	S 243	Cl 2500	Ar
K 173	Ca 27.5	Sc 0.17	Ti 0.32	V 1.3	Cr 0.2	Mn 0.04	Fe 0.00015	Co 0.0015	Ni 1.7	Cu 0.04	Zn 0.02	Ga 0.2	Ge 1	As 0.85	Se 2.2	Br 3350	Kr
Rb 120	Sr 109	Y 0.18	Zr —	Nb —	Mo 18.3	Tc —	Ru —	Rh —	Pd —	Ag 0.009	Cd 8.0	In —	Sn 0.013	Sb 2.1	Te —	I 8.0	Xe
Cs 14.5	Ba 0.7	La 0.094	Hf —	Ta —	W 3.3	Re —	Os —	Ir —	Pt —	Au 2.5	Hg 0.14	Tl —	Pb 0.002	Bi —	Po —	At —	Rn
Fr —	Ra —	Ac —															

La 0.094	Ce 0.044	Pr 0.14	Nd 0.11	Pm —	Sm 0.10	Eu 0.10	Gd 0.11	Tb 0.14	Dy 0.15	Ho 0.20	Er 0.22	Tm 0.21	Yb 0.25	Lu 0.22
Ac —	Th 0.0006	Pa —	U 2.7	Np —	Pu —	Am —	Cm —	Bk —	Cf —	Es —	Fm —	Md —	No —	Lu 0.22

"—" no data or no consistent data in the literature

Dependency of K_{SP} on complex stability

In general, the formation of complexes increases the solubility, while the dissociation of complexes decreases it.

The extent to which elements are soluble and thus more mobile is indicated in Table 6. There, the relative enrichment of the elements in sea water compared to river water is depicted in a periodic system. Substances, which are readily soluble and thus highly mobile are enriched in seawater, whereas less mobile and less soluble substances are depleted.

1.1.4.1.2. Saturation index

The logarithm of the quotient of the ion-activity product (IAP) and solubility-product (K_{SP}) is called the saturation index (SI). The IAP is the product of element activities. Analytically determined concentrations have to be transformed to activities considering ionic strength, temperature, and complex formation. The solubility-product is the maximum possible solubility (based on equilibrium solubility data taken from the literature) at the respective water temperature.

$$SI = \log \frac{IAP}{K_{SP}}$$

Eq.(35.)

The saturation index SI indicates, if a solution is in equilibrium, under-saturated or super-saturated with regard to a solid phase. A value of 1 signifies a ten-fold supersaturation, a value of -2 a hundred-fold undersaturation in relation to a certain mineral phase. In practice, equilibrium can be assumed for a range of -0.05 to 0.05. If the determined SI value is below -0.05 the solution is undersaturated with regard to the corresponding mineral, if SI exceeds +0.05 the water is supersaturated with respect to this mineral. Note that supersaturation is not automatically equivalent to precipitation. If precipitation kinetics are slow solutions can remain supersaturated with regard to mineral phases for very long times.

1.1.4.1.3. Limiting mineral phases

Some elements in aquatic systems exist only at low concentrations (µg/L range) in spite of readily soluble minerals. This phenomenon is not always caused by a generally low occurrence of the respective element in the earth crust mineral as for instance with uranium. Possible limiting factors are the formation of new minerals, co-precipitation, incongruent solutions, and the formation of solid-solution minerals (i.e. mixed minerals).

Formation of new minerals

For example Ca^{2+}, in the presence of SO_4^{2-} or CO_3^{2} can be precipitated as gypsum or calcite, respectively. A limiting mineral phase for Ba^{2+} in the presence of sulfate is $BaSO_4$ (barite). If, for instance, a sulfate-containing groundwater is mixed with a $BaCl_2$-containing groundwater, barite becomes the limiting phase and is precipitated until the saturation index for barite attains the value of zero.

Co-precipitation

For elements like radium, arsenic, beryllium, thallium, molybdenum, and many others, not only the low solubility of the related minerals but also the co-precipitation or adsorption with other minerals, plays an important role. For instance radium is co-precipitated with iron hydroxides or barium sulfate. In the case of iron co-precipitation, radium mobility is also determined by the redox chemistry, since iron is a redox-sensitive ion, which only forms iron oxyhydroxides under oxidizing conditions. Thus, radium, even though it forms only divalent species, mimics the behavior of a redox-sensitive element.

Incongruent solutions

Solution processes, in which one mineral dissolves, while another mineral inevitably precipitates, are called incongruent. When e.g. adding dolomite to water in equilibrium with calcite (SI = 0), dolomite dissolves until equilibrium for dolomite is established. That leads consequently to an increase of Ca, Mg, and C concentrations in water, which in turn inevitably causes super-saturation with respect to calcite and thus precipitation of calcite.

Solid solutions

The examination of naturally occurring minerals shows that pure mineral phases are rare. In particular they frequently contain trace elements as well as common elements. Classic examples of solid-solution minerals are dolomite or the calcite/rhodocrosite, calcite/strontianite, and calcite/otavite systems.

For these carbonates, the calculation of the saturation index gets more difficult. If, for instance, one considers the calcite/strontianite system, the solubility of both mineral phases is estimated by:

$$K_{calcite} = \frac{\{Ca^{2+}\} \cdot \{CO_3^{2-}\}}{\{CaCO_3\}_s}$$

Eq.(36.)

and

$$K_{strontianite} = \frac{\{Sr^{2+}\} \cdot \{CO_3^{2-}\}}{\{SrCO_3\}_s}$$

Eq.(37.)

Assuming a solid-solution mineral made up from a mixture of these two minerals, the conversion of the equations results in:

$$\frac{\{Sr^{2+}\}}{\{Ca^{2+}\}} = \frac{K_{strontianite} \cdot \{SrCO_3\}_s}{K_{calcite} \cdot \{CaCO_3\}_s}$$

$$\text{Eq.(38.)}$$

That means that a certain activity ratio of Sr and Ca in aqueous solution is associated with a certain activity ratio in the minerals. Introducing, analogously to the non-ideal behavior of the activity coefficient of the aquatic species, a specific correction factor $f_{calcite}$ and $f_{strontianite}$ for the activity, leads to the following equation:

$$\frac{K_{strontianite} \cdot f_{strontianite}}{K_{calcite} \cdot f_{calcite}} = \frac{\{Sr\} \cdot X_{calcite}}{\{Ca\} \cdot X_{strontianite}}$$

$$\text{Eq.(39.)}$$

where X is the molar proportion in the solid-solution mineral. In the simplest case, the ratio of both activity coefficients can be combined in order to obtain a distribution coefficient. The latter can be experimentally determined by semi-empirical approximation in the laboratory.

Using the solubility-product constants for calcite and strontianite and assuming a calcium activity of 1.6 mmol/L, a distribution coefficient of 0.8 for strontium and 0.98 for calcite, and a ratio of 50:1 (=0.02) in the solid-solution mineral, the following equation gives the activity of strontium:

$$\{Sr\} = \frac{K_{strontianite} \cdot f_{strontianite} \cdot X_{strontianite} \cdot \{Ca\}}{K_{calcite} \cdot f_{calcite} \cdot X_{calcite}}$$

$$\text{Eq.(40.)}$$

$$= \frac{10^{-9.271} \cdot 0.8 \cdot 0.02 \cdot 1.6 \cdot 10^{-3}}{10^{-8.48} \cdot 0.98} = 4.2 \cdot 10^{-6} \, mol/l$$

If strontianite is assumed to be the limiting phase, significantly more strontium (activity approx. $2.4 \cdot 10^{-4}$ mol/L) could be dissolved compared to that of the solid-solution mineral phase.

This example shows a tendency with solid-solution minerals. There is a supersaturation or an equilibrium regarding the solid-solution minerals but an undersaturation with respect to the pure mineral phases, i.e. the solid-solution mineral is formed but none of the pure mineral phases. The importance of this process depends upon the values of the activity coefficient of the solid-solution component.

For the calculation of solid-solution mineral behavior, two conceptual models may be used: the end-member model (any mixing of two or more phases) and the site-mixing model (substituting elements can replace certain elements only at certain sites within the crystal structure).

For some elements, limiting phases (pure minerals and solid-solution minerals) are irrelevant. Thus, there are no limiting mineral phases for Na or B under natural groundwater conditions. Sorption on organic matter (humic and fulvic acids), on clay minerals or iron oxyhydroxides as well as cation exchange may be limiting factors instead of mineral formation. This issue will be addressed in the following.

1.1.4.2 Sorption

The term sorption combines matrix sorption and surface sorption. Matrix sorption can be described as the relatively unspecific exchange of constituents contained in water into the porous matrix of a rock ("absorption"). Surface sorption is understood to be the accretion of atoms or molecules of solutes, gases or vapor at a phase boundary ("adsorption"). In the following only surface sorption will be addressed more thoroughly.

Surface sorption may occur by physical binding forces (van de Waals forces, physisorption), by chemical bonding (Coulomb forces) or by hydrogen bonding (chemisorption). A complete saturation of all free bonds at the defined surface sites is possible involving specific lattice sites and/or functional groups (surface complexation, chapter 1.1.4.2.3). While physisorption is reversible in most cases, remobilization of constituents bound by chemisorption is difficult. Ion exchange is based on electrostatic interactions between differently charged molecules.

1.1.4.2.1.   Hydrophobic/hydrophilic substances

Rocks may be hydrophobic or hydrophilic and this property is closely related to the extent of sorption. In contrast to hydrophilic materials, hydrophobic substances show no free valences or electrostatic charges at their surfaces. Hence, neither hydrated water molecules nor dissolved species can be bound to the surface and in the extreme case, surface wetting with aqueous solution could be completely suppressed.

1.1.4.2.2.   Ion exchange

The ability of solid substances to exchange cations or anions with cation or anions in aqueous solution is called ion-exchange capacity. In natural systems anions are exchanged very rarely, in contrast to cations, which exchange more readily forming a succession of decreasing intensity: $Ba^{2+} > Sr^{2+} > Ca^{2+} > Mg^{2+} > Be^{2+}$ and $Cs^+ > K^+ > Na^+ > Li^+$. Generally, multivalent ions (Ca^{2+}) are more strongly bound than monovalent ions (Na^+), yet the selectivity decreases with increasing ionic strength (Stumm and Morgan, 1996). Large ions like Ra^{2+} or Cs^+ as well as small ions like Li^+ or Be^{2+} are merely exchanged to a lower extent. The H+ proton, having a high charge density and small diameter, is an exception and is preferentially absorbed.

Moreover, the strength of the binding depends on the respective sorbent, as Table 7 shows for some metals. The comparison of the relative binding strength is based on the pH, at which 50% of the metals are absorbed ($pH_{50\%}$). The lower this pH value, the stronger this distinct metal is bound to the sorbent, as for instance with Fe-oxides: Pb ($pH_{50\%} = 3.1$) > Cu ($pH_{50\%} = 4.4$) > Zn ($pH_{50\%} = 5.4$) > Ni ($pH_{50\%} = 5.6$) > Cd ($pH_{50\%} = 5.8$) > Co ($pH_{50\%} = 6.0$) > Mn ($pH_{50\%} = 7.8$) (Scheffer and Schachtschabel 1982).

Table 7   Relative binding strength of metals on different sorbents (after Bunzl et al. 1976)

Substance	Relative binding strength
Clay minerals, zeolites	Cu>Pb>Ni>Zn>Hg>Cd
Fe, Mn-oxides and –hydroxides	Pb>Cr=Cu>Zn>Ni>Cd>Co>Mn
Organic matters (in general)	Pb>Cu>Ni>Co>Cd>Zn=Fe>Mn
Humic- and Fulvic acids	Pb>Cu=Zn=Fe
Peat	Cu>Pb>Zn>Cd
degraded peat	Cu>Cd>Zn>Pb>Mn

Corresponding to the respective sorbent, ion exchange capacity additionally depends on the pH value (Table 8).

Table 8   Cation exchange capacity at pH 7 and their pH dependency (after Langmuir 1997)

Substance	CEC (meq/100g)	pH dependency
Clay minerals		
Kaolinite	3-15	high
Illite and Chlorite	10-40	low
Smectite Montmorrilonite	80-150	rare or non existent
Vermiculite	100-150	negligible
Zeolites	100-400	negligible
Mn (IV) and Fe (III) Oxyhydroxides	100-740	high
Humic matter	100-500	high
synthetic cation exchangers	290-1020	low

Fig. 10 shows the pH-dependent sorption of metal cations; Fig. 11 the same for selected anions on iron hydroxide.

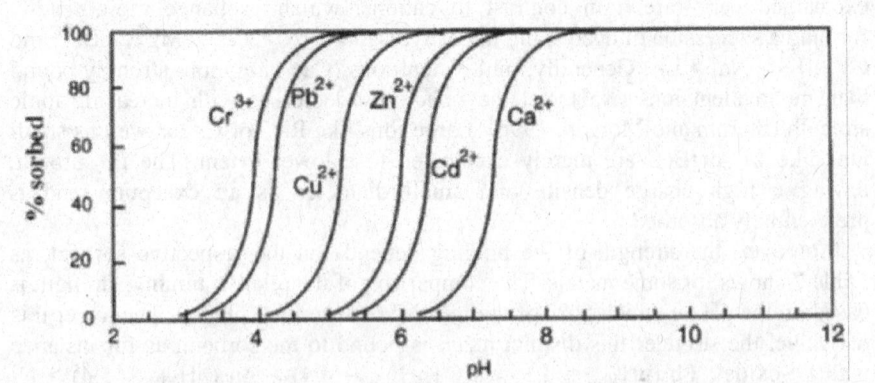

Fig. 10   pH-dependent sorption of metal cations on iron hydroxide (after Drever 1997)

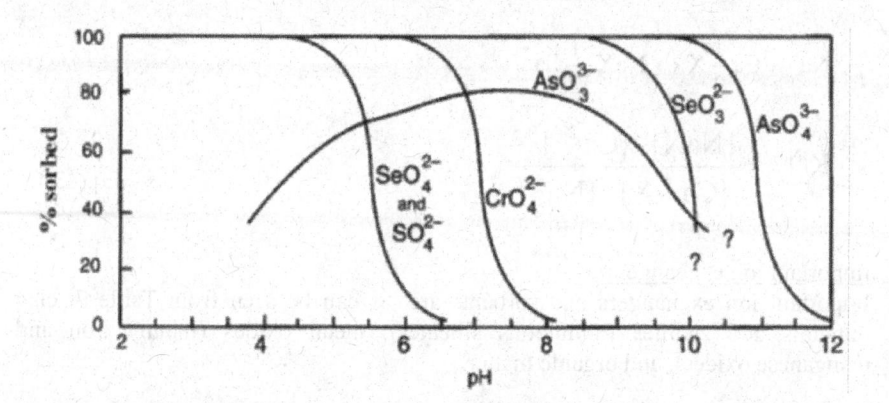

Fig. 11   pH-dependent sorption of anions on iron hydroxide (after Drever 1997)

Description of the ion exchange using the mass-action law
Assuming a complete reversibility of sorption, the ion exchange can be described
by the mass-action law. The advantage of this approach is that virtually any
number of species can interact at the surface of a mineral.

$$A^+ + B^+R^- \leftrightarrow A^+R^- + B^+$$

$$K_B^A = \frac{\{A^+R^-\} \cdot \{B^+\}}{\{A^+\} \cdot \{B^+R^-\}} = \frac{\{A^+R^-\}/\{A^+\}}{\{B^+R^-\}/\{B^+\}} \qquad \text{Eq.(41.)}$$

with A^+, B^+ monovalent ions
            R= exchanger

K_x is the selectivity coefficient and is considered here as an equilibrium constant,
even though, in contrast to complexation constants or dissociation constants, it
depends not only on pressure, temperature, and ionic strength, but also on the
respective solid phase with its specific properties of the inner and outer surfaces.
Although to a lesser extent, it also depends on they way the reaction is written.
    Thus, the exchange of sodium for calcium can be written as follows:

$$Na^+ + \frac{1}{2}CaX_2 \leftrightarrow NaX \cdot \frac{1}{2}Ca^{2+}$$

$$K_{Ca}^{Na} = \frac{\{NaX\}\{Ca^{2+}\}^{0.5}}{\{CaX_2\} \cdot \{Na^+\}} \qquad \text{Eq.(42.)}$$

This expression is called the Gaines-Thomas convention (Gaines and Thomas
1953). Using the molar concentration instead the reaction description is called
Vanselow convention (Vanselow 1932). Gapon (1933) proposed the following
form:

$$Na^+ + Ca\frac{1}{2}X \leftrightarrow NaX \cdot \frac{1}{2}Ca^{2+}$$

$$K_{Ca}^{Na} = \frac{\{NaX\} \cdot \{Ca^{2+}\}^{0.5}}{\{Ca\frac{1}{2}X\} \cdot \{Na^+\}}$$

Eq.(43.)

Important ion exchangers

Important ion exchangers and sorbents are, as can be seen from Table 7, clay minerals and zeolites (aluminous silicates), metal oxides (mainly iron and manganese oxides), and organic matter.

- Clay minerals consist of 1 to n layers of Si-O tetrahedrons and of 1 to n layers of aluminum hydroxide octahedral layers (gibbsite). Al very often replaces Si in the tetrahedral layer as well as Mg does for Al in the octahedral layer.
- As ion exchanger, zeolites play an important role in volcanic rocks and marine sediments.
- At the end of the weathering process, often iron and manganese oxides form. Manganese oxides usually form an octahedral arrangement resembling gibbsite. Hematite (Fe_2O_3) and goethite (FeOOH) also show a similar octahedral structure.
- Following Schnitzer (1986) 70 to 80% of organic matter are humic substances. These are condensed polymers composed of aromatic and aliphatic components, which form through the decomposition of living cells of plants and animals by microorganisms. Humic substances are hydrophilic, of dark color and show molecular masses of some hundreds to many thousands. They show widely differing functional groups being able to interact with metal ions. Humic substances (refractory organic acids) can be subdivided into humic and fulvic acids. Humic acids are soluble under alkaline conditions and precipitate under acid conditions. Fulvic acids are soluble under basic and acidic conditions.

Ion exchange or sorption can also occur on colloids, since colloids possess an electric surface charge, at which ions can be exchanged or sorptively bound. The proportion of colloids not caught in small pores preferentially utilizes larger pores, thus travelling faster than the water molecules in the aquifer (size-exclusion effect). That is why the colloid-bound contaminant transport is of such special importance.

Furthermore, there are synthetic ion exchangers, which are important for water desalination. They are composed of organic macromolecules. Their porous network, made up from hydrocarbon chains, may bind negatively charged groups (cation exchanger) or positively charged groups (anion exchanger). Cation exchangers are based mostly on sulfo-acidic groups with an organic group, anion exchangers are based on substituted ammonium groups with an organic group.

Surface charges

The cation-exchange capacity of clay minerals is in a range of 3 to 150 meq/100g (Table 8). These extremely high exchange capacities are due to two physical reasons:

- extremely large surfaces
- electric charges of the surfaces

These electric charges can be subdivided into:

- permanent charges
- variable charges

Permanent surface charges can be related to the substitution of metals within the crystal lattice (isomorphism). Since usually metals are replaced by other metals of a lower charge, this results in an overall deficit in positive charge for the crystal. A negative potential forms at the surface causing positively charged metals to sorb. The surface charges of clay minerals can be predominantly related to isomorphism, therefore they are permanent to a great portion. However, this is not true for all clay minerals; for kaolinite the permanent charge is less than 50% (Bohn et al., 1979).

Besides the permanent charge, there are variable surface charges, which depend on the pH of the water. They arise from protonation and deprotonation of functional groups at the surface. Under acid conditions, protons are sorbed on the functional groups that cause an overall positive charge on the surface. Thus the mineral or parts of it behave as an anion exchanger. With high pH, the oxygen atoms of the functional groups stay deprotonized and the mineral, or parts of it, shows an overall negative charge; therefore cations can be sorbed.

For every mineral there is a pH value at which the positive charge caused by protonization equals the negative charge caused by deprotonization, so that the overall charge is zero. This pH is called the pH_{PZC} (Point of Zero Charge). If only deprotonization and protonization have an influence on the surface charge this value is called ZPNPC (zero point of net proton charge) or IEP (iso-electric point). This point is around pH 2.0 for quartz, around pH 3.5 for kaolinite, for goethite, magnetite, and hematite approximately between pH 6 and 7, and for corundum around pH 9.1 (Drever 1997). Fig. 12 shows the pH-dependent sorption behavior of iron hydroxide surfaces. The overall potential of the pH-dependent surface charge does not depend on the ionic strength of water.

Natural systems are a mixture of minerals with permanent and variable surface charge. Fig. 13 shows the general behavior in relation to anion and cation sorption. At values exceeding pH 3 the anion exchange capacity decreases considerably. Up to pH 5 the cation exchange capacity is constant, rising extremely at higher values.

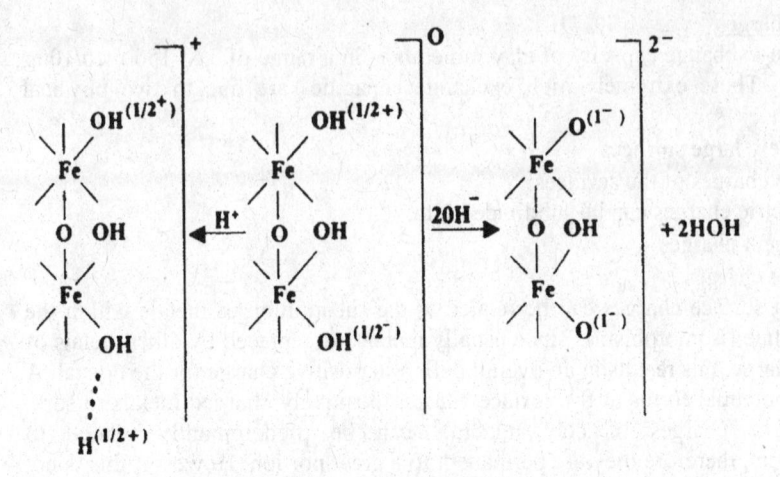

Fig. 12   pH-dependent sorption behavior of iron hydroxide surfaces after binding of H⁺ under acid and OH⁻ ions under alkaline conditions (after Sparks 1986).

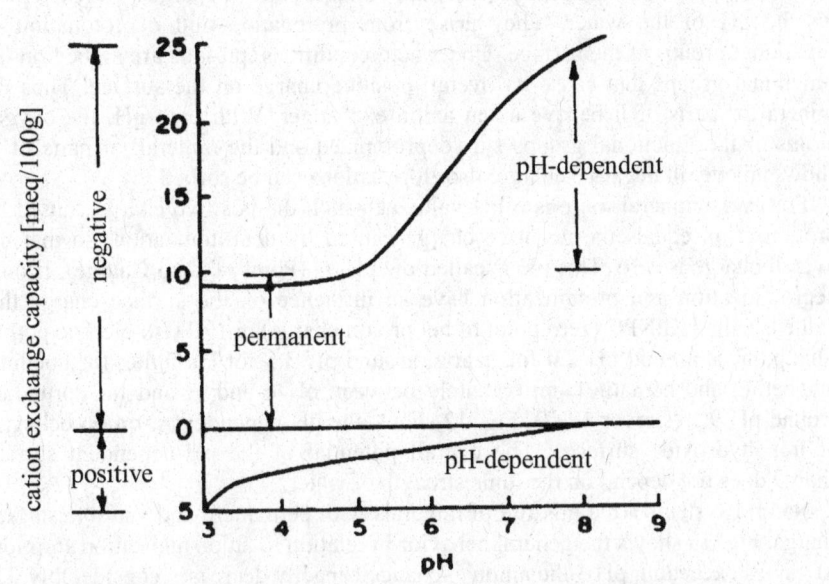

Fig. 13   Cation and anion exchange behavior of minerals as a function of pH (after Bohn et al. 1979); "negative" and "positive" relates to the charge of the surfaces, so that "negatives" are cation exchanger and "positives" anion exchanger.

1.1.4.2.3.   Mathematical description of the sorption

Numerous equations are used to describe experimental data obtained for interactions between liquid and solid phases. These equations range from simple empirical equations (sorption isotherms) to complicated mechanistic models based

representation drawings to manufacture a building and its components, like CNC fabrication technologies, the use of robots requires a different type of computing which "generates the design out of the fabrication parameters and the sequential fabrication steps" [Bonwetsch 2012].

To change the shape of a material we may use formative technologies, which apply forces to the material in order to get the desired final shape. Bending, extrusion, thermoforming and molding are some examples of formative technologies [Pupo et al 2009: 441]. Due to the high price of machinery the use of this technology in architectural reearch is still residual and mostly industry-oriented.

Along with these new ways of creating complex surfaces, patterns or forms there is an increasing interest in research in construction materials. This research is focussed on concepts like biomimetics, morphogenesis, generative systems, complexity and emergence among others. The work carried out by Neri Oxman at MIT within the initiative "Materialecology" is a great example of research in computational form-generation inspired by nature [Oxman 2010].

Kolarevic and Klinger [2008:3] emphasise the fact that projects growing out of research in digital fabrication exploring mathematics logics of surface modelling are dependent on software that is entirely surface-oriented in its underlying mathematics. Decisions made during design, prototyping, fabrication and assembly rely on codes, scripts, parameters, operating systems and software. This situation creates the need for teams with multidisciplinary expertise and different skills, from IT to architecture, design, material engineering, biology and mathematics, among others.

The Internet has considerably increased the possibilities of collaboration between different realities, expertise, cultures, etc. A paradigm emerged with the awareness that knowledge relies on the interdisciplinary and international collaboration between networks of experts, researchers, practitioners, students and ordinary people. These networks are usually based in online blogs, chats or forums that act as repositories of experience and code.

According to Neil Gershenfeld, the creator of MIT's FabLab network, fabrication laboratories are places to make (almost) anything anywhere [Gershenfeld 2005]. They empower people, especially in developing communities, to design and create tools to solve local problems.

Digital design and fabrication technologies combined with customized prefabrication may create innovation in materials' manufacturing industries because of the possibilities for customization and the fast design-to-production process [Mitchell and McCullough 1994]. Digital technologies allow for a further embedding of information within design processes, including information such as material properties, environmental parameters, user's data, time, construction regulations and construction methods. A process of design based on the use of several local parameters results on a customization of the final product.

The use of digital technologies in the manufacturing and building industries connects designers to manufacturers, enabling an optimized, more efficient process which does away with the traditional constraints of industrial standardization [Sheil 2013]. According to Sheil, today's bespoke architecture is a customized architecture produced by mechanical processes which takes account of specific parameters to mass-produce unique pieces.

Several research projects oriented to industry innovation have been completed during the past years. Examples are: rethinking the use of cork in architecture by José Pedro Sousa [2010], rethinking ceramic tiles by Duarte and Caldas [2005], translucent metal panels by Patterns [Kolarevic and Kevin 2008: 49]; construction system of reinforced concrete panels by Alvarado [2009]; new directions for stone by Studio Gang architects [Kolarevic and Kevin 2008: 81], among others.

Although CNC manufacturing allows the production of components in a very accurate way, the process of assembly is usually carried out through by traditional manual methods which means that typical tolerances are re-introduced into the new manufactured objects [Shelden 2002: 47]. The use of robotics in the assembly process reduces or removes this lack of accuracy and allows a complete correspondence between the design and the on-site final product. The work of Gramazio and Kohler and their team at ETH emphasizes construction as an integral part of architectural design by controlling and manipulating the building process with robots [Bonwetsch 2012].

In architectural practice these digital technologies are being used at several scales of intervention. Besides Frank Gehry's innovations in almost all aspects of digital design and fabrication, other practitioners are exploring technologies in some very interesting ways. Offices like Jean Nouvel, Zaha Hadid, Foster and Partners and Amanda Levete architects, are among the best practices in these areas.

In the Galaxy Soho project Zaha Hadid explores complex geometries and their translation into construction through design and fabrication technologies. The Louvre Abu Dhabi project by Jean Nouvel used digital design and fabrication technologies to test strategies for extreme environment conditions. Similar explorations were carried out by Enrique Ruiz-Geli and Cloud 9 with projects like Villa Nurbs which explores the limits of NURBS technology as well as CAD/CAM and Media-ICT with its CAD/CAM manufactured facade. Foster and Partners, in collaboration with Loughborough University, have recently investigated large scale free-form construction using additive technologies [Kestelier 2011].

None of the above practices in architecture were driven by cost constraints and were only made possible because they were unique projects for private real estate developers. In contrast, there are several initiatives in designing customized housing. Within these projects we highlight the Instant House developed in MIT [Sass and Botha 2006] and FACIT houses designed by Bruce Bell. The Instant House is an experimental project combining prefabrication with digital fabrication with the goal of creating customized housing made of wood derivates for emerging communities. FACIT houses are digitally designed and manufactured on a CNC router using plywood. The goal is to create a cost-effective, bespoke house using digital technology.

Mark Burry used digital design and fabrication methods to complete the Sagrada Familia, Gaudí's unfinished masterpiece in Barcelona. Burry's pioneering methods involve parametric design, material computation and high-tech digital fabrication in stone and concrete.

Diez [2012] states that today's architecture is full of unique, inimitable and iconic creations which try to make a statement in the territory. This approach to the use of CAD/CAE/CAM technologies enhances the least desirable attributes of these technologies and constitutes a serious error. The point of view of Amanda Levete

Architects [2011] on using digital technologies is that the great freedom in form-making comes with great responsibility.

Symposium Digital Fabrication – A State of The Art

The "Symposium Digital Fabrication – a State of the Art" was held at ISCTE-IUL on the 9-10 September 2011. The aim of this symposium was to define the state of the art of digital fabrication in architecture and other areas such as design, fine arts and building construction.

The fusion of architecture, computing and manufacturing enabled by the use of CAD/CAM technologies has, in recent years, broadened the scope of research in architecture and transformed traditional approaches to the architectural issues.

At this symposium we challenged the participants to explore how computing can improve the design and performance of architecture, how the experience and perception of space can be enriched through the integration of emerging technologies, how the shape can be generated parametrically and evolve and, above all, how to take advantage of digital modeling and manufacturing to produce differentiated physical objects.

Digital manufacturing is still poorly understood and used by architects in Portugal, with only a small number of researchers actively involved. Given this background, the symposium aimed at a wide dissemination among non-specialists and a specialized audience in architecture, fine arts, engineering and construction, in order to make evident the potential of this approach, especially in regard to the architectural object.

The presence of some of the most respected international experts in the field of manufacturing and digital media allowed a reflection and debate on the role of these technologies in current practice and education.

Two decades after the first attempts at utilizing digital fabrication in architecture, this symposium provided an account of today's progress and experimentation across the globe. Works from Europe with Bob Sheil, Tobias Bonwetsch and Tomas Diez, from Brazil with Gabriela Celani and from the USA with Kevin Klinger were shown and debated. The five talks by these researchers and practitioners were complemented by a workshop conducted by the Portuguese researcher José Pedro Sousa. In this workshop the participants were able to experiment with CAD/CAM technologies from concept creation to final production using Portuguese cork by Amorim and color medium-density fibreboard (MDF) by Valchromat on a 3-axis CNC router.

Unlike other conferences related to this theme, the Symposium Digital Fabrication was sponsored by Portuguese material manufacturing industries, who stressed their belief that digital fabrication and digital technologies are great tools for optimizing the materialization of architecture. By working together with these industries, architectural research can be done in close collaboration with the manufacturers and informs and is informed directly by the actual fabricators.

Keynote speakers presented widely representative works on Digital Fabrication. Kevin Klinger opened the symposium by providing an insight into how to see digital fabrication and the worldwide connectivity in the process of designing architecture. Klinger pointed out three major aspects associated to the new paradigm of architecture: the design-through-production system; the possibility of producing locally something that has been designed or informed by the global network; the new paradigm in which form follows performance. The need to form partnerships between academia and the manufacturers of building materials was one of the most discussed aspects among the speakers and this was stressed several times by Klinger. According to him, this partnership allows the testing of new products in real environments and increases the quality of industrial products. Klinger challenged academia to produce knowledge that may have real impact in society.

Bob Sheil addressed both the role of digital technologies in today's education and in architectural practice. In his talk, "Manufacturing Bespoke Architecture", Sheil noted that although digital technologies are emerging in architectural degree programs we have to work hard to integrate this new tools in the creative process of students. Sheil pointed

out that architecture students still see CAD and CAM technologies solely as a tool to visualize and produce scale models. The possibility of using digital manufacturing technologies to test and optimize the digital models enables architects to design architectural solutions that can be proven to be adequate prior to their final production and assembly. Regarding this issue Sheil highlighted the need to "make" and to "build" during architectural degree courses using 1:1 scale models.

Tomas Diez, from Spain, talked about "Personal Fabrication: Fab Labs as a Platform for Change, From Microcontrollers to Cities". Diez showed several projects of the Insitute for Advanced Architecture of Catalonia in which he was involved, and particularly projects in urban areas which involve the interaction between man and machine.

The Brazilian experience was presented by Gabriela Celani from the University of Campinas, who focused on education. In her paper presented here, entitled 'Digital Fabrication Laboratories: Pedagogy and Impacts on Architectural Education" Celani reviewed the teaching of architecture throughout the ages to promote a reflection on the meaning of digital fabrication laboratories in schools of architecture today. She outlines three possible lines of action for a laboratory – research, applied development, and education – and concludes that most schools of architecture have sought to reconcile the three in order to explore all areas, thus resulting in higher productivity in comparison with cases of isolated approaches.

Celani concluded by saying that digital manufacturing is revolutionizing the process of designing and manufacturing in architecture, a revolution in which the paradigm of form as the driving force of the original architectural idea is being matched by the materiality and the structure.

The final lecture of the symposium was given by Tobias Bonwetsch from ETH Zurich, Switzerland, where he serves as Research Fellow in the field of Architecture and Digital Fabrication.

Bonwetsch's lecture was entitled "Tailoring fabrication processes" and focussed mainly on the application of robotics in the assembly of contruction components. He presented the work developed at ETH Zurich, which, employing much more powerful equipment such as industrial robots in adition to the milling machines, laser cutters and 3D printers, has gone one step further in the application of the technology of design and manufacturing of digital architecture. Bonwetsch showed how robotics can be incorporated into the design process and emphasized the various operations that a single robot can accomplish just by changing the tool with which it is equipped, such as cutting or adding material, making it much more versatile than a machine with only one function. In addition to the various possibilities of operation on the material, robots can also assemble components, something that no other digital equipment can do, and do it with high precision, resulting in complex assemblies that are very difficult, if not impossible to do by hand.

In his lecture Bonwetsch revealed a series of new possibilities in the field of digital manufacturing and was asked several times by the audience to expand on the capabilities of robots and human-machine interaction.

In the afternoons of both symposium days a workshop was given by José Pedro Sousa of FAUP / DARQ / OpoLab with the title "Digital Explorations > Material Realities". Sousa is an architect, researcher and teacher interested in exploring the conceptual and

material possibilities that arise from the use of technologies for digital design and manufacturing. This researcher recently defended his doctoral thesis, entitled "From Digital to Material: Rethinking Architecture in Cork through the use of CAD/CAM Technologies" [2010], which explores the potential use of cork in architecture through digital technologies.

The workshop held during this symposium was intended to provide the public, composed of students, architects, designers, plastic artists and other professionals with an opportunity to experiment and try new methods of approach to the themes discussed in the cycle of conferences. The two workshop sessions were held along with the two-day event, allowing, in the first session, the first contact with three-dimensional modeling techniques using the Rhinoceros software (fig. 1) and, in the second session, experimentation with digital subtractive manufacturing techniques (fig. 2). The workshop was attended by twenty-five people from several countries so the official language, as in the symposium, was English.

Fig. 1. First day of the workshop, work on 3D modeling with Rhino

Fig. 2. Second day of the workshop, milling cork with a CNC router from Ouplan

Fig. 3. Some results of the workshop using cork from Amorim and color MDF from Valchromat

The works produced in the workshop allowed several issues to emerge. On one hand, complex geometries were explored as well as their transition from digital to material. On the other hand, digital manufacturing was explored as an aid to designing, in the sense that real-time production of the model allows a rapid evaluation, quick alterations and re-fabrication. Other aspects covered during the workshop were the possibilities of working with cork and MDF both in terms of the cutting and surface cutting texture development (fig. 3).

During the Symposium, the SUTURE installation by New York-based alUm Studio – Carla Leitão and Ed Keller – was exhibited (figs. 4 and 5). SUTURE is an expanded cinema installation which "proposes a new architectural body created through event, gesture and temporality informed by a contemporary reconsideration of cinematic and architectural affect" [Leitão and Keller 2011]. The installation used digital video which, according to the authors, acts as a critical lens and interface to expose the relations between humans and their environment.

The SUTURE installation inhabited the ISCTE-IUL's large, double-height exhibition room and formed a mandatory path between the symposium's conference venue and the workshop room. The path started at a large scenic ramp which led the visitor down to the exhibition room, allowing a wide view into the entire space. As visitors encounter the installation at the ramp, their gestures mediated between abstract and real space cinematic and haptic events. Motion and presence sensors were placed in the columns along the ramp providing interaction points and encouraging visitors to create new signal paths and new cycles in the space. In addition, three networked computers, three video projections and a soundscape form the layers of event, which accumulate and mutate according to rules of self-organization in the network [Leitão and Keller 2011].

Figs. 4-5. SUTURE installation by alUm Studio, Carla Leitão and Ed Keller, Programming by George Showman. Photos by João Morgado - Fotografia de Arquitectura

Digital fabrication at ISCTE-IUL

Digital fabrication is a recent area of expertise in ISCTE-University Institute of Lisbon. It was the result of a number of contributing aspects. First we will briefly explain the set up and the inherent development strategy of the laboratory. Second, an overview of ongoing or recent research projects is presented.

ISCTE-IUL Vitruvius FabLab set up

The creation of a digital fabrication laboratory at ISCTE-IUL was based on three main assumptions. The first starting point was the recognition of a great potential inside ISCTE-IUL in taking advantage of latent synergies. With regard to digital fabrication, our university has other strong scientific groups besides architecture that were interested in the laboratory, namely computer sciences, entrepreneurship and society and technology. One other assumption was the desire and the need to further develop a clear differentiation within Portuguese higher education in architecture. To follow this objective, specific high-level scientific competence was needed. Therefore, the third assumption was the fact that the Department of Architecture and Urbanism had, at the time, two teachers finishing Ph.D. research in this field of expertise. Moreover, other existing specific expertise included construction materials and structural engineering.

From these premises, a fundamental question arose. What should the scope of the laboratory be? This issue has been addressed over the past few years by scholars and practitioners. At the "Symposium Digital Fabrication: a State of the Art", Gabriela Celani presented a very interesting view about how digital fabrication laboratories in academic environment serve three main goals: research, development and education. Celani argues that there is a challenge in fulfilling these objectives keeping track of the need to incorporate in architecture curricula the ever new possibilities brought by digital fabrication.

Research is a fundamental activity of the academic environment, which means that a university laboratory has to comply with the respective requirements. On the other hand, research in architecture has to focus, at some point, on solving specific practical problems. Accordingly, an academic laboratory in a school of architecture should also allow for practical problem solving. In a digital fabrication facility, this means, among other things, the development of construction materials and systems. This task is better undertaken in partnership with the industry corresponding to the development facet. The association with industry may also allow for a financial turnover, which is quite relevant in ISCTE-IUL's strategy to face the actual economic environment by promoting entrepreneurship.

With regard to education, there is indeed a huge challenge, at least when considering the overall main objective of giving students effective competencies in digital fabrication related to architectural design. First of all, it should be noted that it is not difficult to capture students' interest in these technologies, since the outputs from digital simulation and fabrication are usually visually impressive. There is a sense of the new and avant-garde that stimulates young people's restless minds. Work done with students exclusively within digital fabrication matters is often very interesting. These kinds of results are obtained in specific courses intended as an introduction to the concepts, software tools and the machinery. However, things become more complex when there is an attempt to go further in the use of these tools.

Incorporating digital fabrication methods and potential in the process of architectural design requires more than just knowing how to use its tools. It is above all a matter of attitude. The main courses in architectural design are usually coordinated by practitioners that do not have a specific experience in using digital fabrication as a tool that informs the design process. The consequence is that students find it difficult to dialog with their design teachers when these technologies are an important part of their architectural proposal and not just a representation tool.

This same problem has been noted with other types of knowledge and expertise. For instance, the use of life-cycle or energy-related issues as a justification for a given design result was (and to some extent still is) often disregarded by some architectural design teachers. This may be in part explained via the *beaux-arts* approach to architecture which has traditionally informed architectural curricula in some countries, namely Portugal. Functional requirements, technological issues, building physics (just to mention a few examples) are often considered questions to be solved with the support of engineers at some point, late in the design process. This means, within the *beaux-arts* approach, that those are not issues that inform architecture but rather problems that complicate the production of architecture.

The integration of digital fabrication and related technologies in architectural education is reliant on another challenge. Unlike the above mentioned issues, to use digital fabrication to its fullest potential means to complement the architectural design process. For example, thermal comfort should inform the idealization of a building from its very beginning because it has consequences on the form, volume, glazing, materials, etc., but thermal comfort is not a design tool. Digital fabrication may inform the generation itself of the building, being a significant part of the design process. This assumption implies that the teaching of architecture, if willing to use these tools, needs to assume new methods.

How to drive this shift is then a major question. The answer is not unique and depends on several conditions. In ISCTE-IUL a two-way strategy has been defined and is now in its implementation phase. Two main criteria form this strategy: encourage and involve graduate students and produce high-level research. The rational is that high-level outcomes will support the explanation of the potential of digital fabrication in architecture and basic competencies among lower-level students are essential for future exploration of that potential.

The first criteria – teaching of basic competencies – has been achieved up to now through optional courses where students learn the basics of shape grammars, auto-LISP programming and digital fabrication. The second criteria – high-level research – is being implemented with ongoing multidisciplinary research projects and through a one-year post-graduate learning programme called Digital Architecture (in partnership with the Faculty of Architecture, Oporto University).

The implementation of this strategy is also based on the establishment of several internal and external partnerships. Internally, ISCTE-IUL specialists in computer sciences are working with architects to form the core team of ongoing research. The main subjects addressed by these colleagues are programming, shape grammars, multi-agent expert system shells and augmented reality. ISCTE-IUL is also one of the partners of the science centre *Ciência Viva do Lousal* where a CAVE Hollowspace is installed. This facility is used within the Digital Architecture programme to produce 3D simulation outputs by students and teachers.

External partnerships include Portuguese manufacturers of building materials, an authorized Rhino Centre, other Fab Labs and the Centre for Research and Studies Art and Multimedia (University of Lisbon).

The described setup of the Vitruvius Digital Fabrication Laboratory (Vitruvius FabLab–IUL) is intended to promote and develop the use of digital fabrication and related technologies in architecture. We foster innovation through the implementation of

the practical theorizing concept [Hagger and McIntyre 2006: 58-59] where new knowledge is informed by the experiment of its formalization. Designing architecture for the future will require ever more reliable and accountable procedures so that the construction and the operation phases are well planned. The optimization underlying this goal may pose a problem for innovation unless there is a means to experiment as much as possible before construction. This experimentation, which validates the knowledge produced, is made possible to a significant extent by digital fabrication.

Research projects

Vitruvius FabLab research is focused on architecture and addresses issues such as design logic form, electronic technology, artificial life, robotics and human computer interaction. Supported by these guidelines two main complementary projects are being developed: Emerg.cities4all and bio-construction through natural matter with computational form generation. We will also dicuss a recent workshop entitled "Discursive Wall – a Living System".

Emerg.cities4all. Emerg.cities4all is an ongoing research project focused on the development of a generative computer-aided planning support system for cities and housing for low-income populations. The system uses a descriptive method – shape grammars – and is based on a direct input strategy targeted to the common user with no digital expertise.

The Emerg.cities4all project arose with the goal of contributing new solutions to the housing problems of the economically underdeveloped countries, which are facing a rapid, uncontrolled urbanization. However, mass homelessness is still a consequence of the rural migration to urban peripheries. The consequence is the increase of informal mass construction settlements. Recent attempts to address these problems seem to have solved just a part of it and have created new issues as a result of the generalized and uncritical approach [Paio, et al. 2011].

The project assumes that it is possible to generate mass housing (creating an urban environment, houses and building solutions) based on the concept of mass personalization. This goal is achieved by creating a generative computer-aided system that develops a set of possible solutions to suit specific parameters related to local conditions. These conditions include climate, material resources, type of family, social characteristics and economic constraints.

Solution generation is based on a descriptive method such as shape grammars [Stiny 1980] as a generational urban settlement and multiplicity housing method, guaranteeing adaptability and evolutionary capabilities. The research is using this tool not only to generate new forms but also to better understand existing settlements through analytical grammars. Within this scope, the proposed shape grammar seeks to analyse how existing slums, *favelas* and *musseques* are generated and what social, cultural and spatial dynamics are involved in their growth [Paio, et al. 2011]. Despite the terrible living conditions, there are reasons to believe that the adaptability and evolutionary characteristics of their houses, as well as their social and spatial relations, have some inherent good qualities. Not having the intention to recreate existing settlements exactly as they stand, the new original grammars will also denote a human approach because they are informed by local humanized logics and specific ways of self spatial organization [Linhares, et al. 2011].

Regarding the feasibility of the generated design, the project will develop a serial construction system whose degree of detailing is dependent on specific partnerships with the construction industry.

The process is conceived on a basis of a multi-agent system, which enables interaction at three different levels: system specialist (builds and expands the shell); shape grammar specialist (builds a system applied to a specific area using the shell); common user (applies the system to create solutions in the specific area). To facilitate interaction with users, the system interface is based on a graphical and symbolic definition of shapes.

Bio-construction through natural matter with computational form generation is a research work based on a Ph.D. thesis being developed at Vitruvius FabLab.

The fundamental question that underlies this project is how to develop an external wall building system that has the ability of self adaptation in response to solar radiation in order to optimize thermal comfort conditions in the building. The project seeks the answer to the above question through biomimicry processes inspired on the evolution of biological self organization.

The ability to mutate requires living system capacities which will depend on specific sensors that will measure solar radiation properties, internal and external air temperature, relative humidity and wind pressure. The combination of this data will trigger the adaptation of the wall surface to change the way solar energy is absorbed and reflected. The goal is the optimization of solar heat gain through the wall considering the adequate thermal comfort strategy in face of the external conditions.

To adapt the wall's external surface requires a construction solution made of small elements combined through complex geometric patterns. This is being conceived by deliberated design criteria using already existing resources such *Variable Property Design* [Oxman 2010] and the *autopoiesis theory* [Maturana and Varela 1971].

Another crucial aspect is the definition of how materials may be used in the construction system. Several possibilities will be explored by analysing material properties such as mechanical strength, dimensional stability, water absorption, thermal conductivity and diffusivity, specific heat capacity, durability, cost and environmental impact. These properties will be integrated to define the most adequate materials to be incorporated in the prototyping phase. Following the definition of the most adequate materials, the assembly system will be addressed. The goal is to obtain a simple construction method that easily incorporates the installation of the sensors and of the movable parts. Prototypes will be fabricated at different scales to evaluate the possible combinations of materials / assembly systems. The ones found most suitable will then be fabricated at scale 1:1 for final validation.

Data collection and interaction criteria will be controlled by open-source platforms. Fabbers and C/C++ programming will be integrated to provide the necessary information to the wall thus creating an autonomous input/output organism.

Apart from the ability to respond to thermal comfort requirements, the wall will also have an interesting architectural feature: it will regenerate itself by changing the appearance of its external surface.

Discursive Wall – a Living System. Discursive Wall – a Living System was an extended workshop that took place at de Vitruvius FabLab–IUL facilities in March 2012. The objective of the workshop was to develop and fabricate an interactive living system

wall prototype. The wall is intended to respond physically to sound interacting with the surrounding environment and establishing a direct dialog with the inhabitants [Oliveira et al. 2012].

The fundamental issue that supports this system is the creation of an architectural living system constantly being designed and re-designed by its inhabitants, minimizing acoustical problems of the space.

The methodology employed to develop the discursive wall includes four steps. The first is to design and fabricate the interaction unit that will define the living system through CAD/CAM/open source process. The second step consists of informing the metabolism and organizing the units of interaction in a closed circular process. This is achieved through a mathematical description method enabling the definition and redefinition of the rules by the computer modelled parametric process. Interactive simulations are produced through the open source platform and C/C^{++}. The third stage is to assure that the circular organization will be respected. To establish this circular organization an open source code based in C/C^{++} is used. Finally, the fourth step is the creation of a response mechanism through the use of sound sensors. These enable the production of different reactions and situations on the uninterrupted living system. Therefore, the discursive wall maintains the capacity of "speaking" to the inhabitants of the room.

Conclusions and future work

The "Symposium Digital Fabrication - A State of Art" provided an opportunity to reflect on digital tools and the gap between the digital fabrication, academia, research and practice. If on the one hand we are aware of paradigms such as design-through-production, the possibility to create and produce *in loco* and the recent "form follows performance", on the other hand students still tend to see CAD/CAM resources as simple tools to produce scale models. Therefore the greatest challenge faced by academia is to produce knowledge through the possibility of using manufacturing technologies, testing it and optimizing their products, establishing the bridge between creators, researchers, the industry and material manufacturers. The incorporation of robotics in the design and manufacturing process, operation material and its use in the assemble components gives us new possibilities for producing architecture.

The digital fabrication technologies methods are now pointing to a new way of conceiving contemporary architecture. Implementing in academia these resources common to the industry, with a strong theoretical and historical basis and providing a straight relationship with cultural identity, will redirect architecture to the basic principles of construction thinking, increasing the potential of research and practice. As in many other areas of knowledge, architectural academia could also be the pioneer in the experimental field, making an intelligent use of digital fabrication tools and their potential. Academy can become the provider of new research, experiments and new tested models that will improve everyday practice with new tools and solutions for future projects. The combination of the right resources, technology, creative minds and a fundamental multidisciplinary knowledge, makes academia's contribution essential to the enrichment of contemporary and future architecture. Robotics, open source and scripting resources are the means that bring us new ways of conceiving the architectural processes. These tools enable us to rethink the traditional design and construction process, by giving us the necessary inputs to test, built and construct anywhere at any moment.

Vitruvius FabLab-IUL is directing its research to the community, trying to solve local problems, looking at the human-machine interaction.

Future work includes an open online platform to follow the concept of sharing within the community. Ongoing projects will be detailed and explained in this online platform, providing all the tools and the experienced knowledge so that anyone, anywhere, may build their own solutions.

One relevant project that we have recently embraced is the "Guardian", an interior façade of a public library that reacts to the excess of noise. The Guardian produces intensive light signs when the library inhabitants produce excessive noise, keeping colors soft when the sound scale is adequate to the space.

At the present moment, three human issues from the past are being addressed again: the 1980s cliché of the human-machine interaction, the 1990s condition of the Skin, and the ancient human necessity to imitate the forms of nature. Recent achievements in these research areas are essentially based in new digital tools. Thus, rethinking these issues, crossing them with our current society needs, with all these emergent conditions of technologies and with its fundamental academic condition is leading us to new projects. To provide a solution for a south facade, we will develop a wall responsive to its own environment. This project aims at solving problems related to sun and natural ventilation through the creation of mutable shadows and mechanisms. In these future works we envision adding higher levels of functionality to the purposes of architecture.

Acknowledgments

The authors would like to acknowledge the sponsors of "Digital Fabrication Symposium - a State of the Art": Calouste Gulbenkian Foundation, Foundation for Science and Technology, Luso American Development Foundation, Amorim Isolamentos and Amorim Cork Composites, Valchromat-Investwood, Ouplan-CNC milling, SimplyRhino and 3D Rhino Portugal.

References

AMANDA LEVETE ARCHITECTS. 2011. Three projects: a comparative study. Pp. 208-215 in *Fabricate: Making Digital Architecture*, R. Glynn and B. Sheil, eds. Toronto: Riverside Architectural Press.

ALVARADO, R. 2009. Constructive Models by Digital Fabrication. *Proceedings of the Sigradi 2009 Conference*, São Paulo: 421-423.

BENYUS, J. 1997. *Biomimicry: Inovation Inspired by Nature*. New York: HarperCollins.

BONWETSCH, T. 2012. Robotic Assembly Processes as a Driver in Architectural Design (Paper presented at Symposium Digital Fabrication: a State of the Art, 9-10 September 2011). *Nexus Network Journal* 14, 3.

BURRY, M. 2011. *Scripting Cultures. Architectural design and programming. AD primers*. London: John Wiley and Sons.

CHENG, N. 2003. Digital curriculums: Effective integration of digital courses. In *Digital technology and architecture-White paper*, J. Bermudez and K. Klinger, eds. ACADIA. See: http://digit-architecture.blogspot.it/2011/03/continue-last-article-2.html.

CELANI, G. 2010. Os Workshops do Sigradi09 e a Fabricação Digital no Brasil. In *Vitruvius*. Available at: http://70.32.107.157/revistas/read/drops/10.030/2114.

———. 2012. Digital Fabrication Laboratories: Pedagogy and Impacts in Architectural Education (Paper presented at Symposium Digital Fabrication: a State of the Art, 9-10 September 2011). *Nexus Network Journal* 14, 3.

DIEZ, T. 2012. Personal Fabrication: Fab Labs as a Platform for Change, from Microcontrollers to Cities (Paper presented at Symposium Digital Fabrication: a State of the Art, 9-10 September 2011). *Nexus Network Journal* 14, 3.

DINIZ, N. 2007. Augmented Membranes. *Proceedings of Ubicomp 2007: Transitive Materials Workshop*, Innsbruck, Austria.

DUARTE, J. 2007. Inserting New Technologies in Undergraduate Architectural Curricula: A Case Study. Pp. 423-430 in *Proceedings of the eCAADe 2007 Conference*, Frankfurt.

DUARTE, J. P. and L. CALDAS. 2005. Fabricating Ceramic Covers: Rethinking Roof Tiles in a Contemporary Context. Pp. 269-276 in *Proceedings of the eCAADe 2005 conference*, Lisbon.

FJELD, P. O. 2008. Changes of paradigms in the basic understanding of architectural research. *ARCC Journal Architectural Research Centers Consortium. Changes of Paradigms* 5, 2, 8: 7.

GRAMAZIO, F., M. KOBLER and S. OESTERLE. 2010. Encoding Material. *AD Architectural Design Magazine: The New Structuralism*, n. 206: 108-115.

GERSHENFELD, N., 2005. *FAB: The Coming Revolution on Your Desktop – From Personal Computers to Personal Fabrication*. Basic Books.

HAGGER, H. and D. MCINTYRE. 2006. *Learning Teaching from Teachers*. Berkshire: Open University Press.

HENSEL, M., A. MENGES and M. WEINSTOCK. 2010. *Emergent Technologies and Design. Towards a Biological Paradigm for Architecture*. London and New York: Routledge.

KESTELIER, X. 2011. Design potential for large-scale additive fabrication. Free-form construction. Pp. 244-249 in *Fabricate: Making Digital Architecture*, R. Glynn and B. Sheil, eds. Toronto: Riverside Architectural Press.

KLINGER, K. 2012. Design-Through-Production Formulations (Paper presented at Symposium Digital Fabrication: a State of the Art, 9-10 September 2011). *Nexus Network Journal* 14, 3.

KOLAREVIK, B. 2001. *Designing and Manufacturing Architecture in the Digital Age*. *Proceedings of the eCAADe 2001 conference*. Helsinki: 117-123

———. 2003. Digital Production/Fabrication. *Digital Technology & Architecture – White Paper ACADIA*: 4.

KOLAVERIC, B. and K. KLINGER. 2008. *Manufacturing Material Effects Rethinking Design and Making in Architecture*. London and New York: Routledge.

LINHARES, B., H. ALARCÃO, L. CARVÃO, P. TOSTE and A. PAIO. 2011. Using Shape Grammar to design ready-made housing for humanized living. Towards a parametric-typological design tool. Pp. 77-79 in *Proceedings of XV Congreso de Sigradi 2011*.

LEITÃO, C. and E. KELLER. 2011. a l *Um Studio: SUTURE*. Catalog of the installation. Lisbon: ISCTE-IUL.

MALÉ-ALEMANY, M., J. AMEIJE and V. VIÑA, V. 2011. (FAB)BOTS Customized robotic devices for design & fabrication. Pp. 40-47 in *Fabricate: Making Digital Architecture*, R. Glynn and B. Sheil, eds. Toronto: Riverside Architectural Press.

MATURANA, H. and F. VARELA. 1980. *Autopoiesis and Cognition. The Realization of the Living*. Boston Studies in the Philosophy of Science, vol. 42. Boston: D. Reidel Publishing Co.

MEREDITH, M. 2008. Never enough (transform, repeat ad nausea). Pp. 4-9 in Pp. 4-9 in *From Control to Design. Parametrical/Algorithmic Architecture*, M. Meredith et al., eds. Barcelona: Actar-D.

MITCHELL, W. and M. MCCULLOUGH. 1994. *Digital Design Media*. London: John Wiley and Sons.

OXMAN, N. 2010. Material-based Design Computation. Ph.D. dissertation, Massachussetts Institute of Technology.

———.2011.Variable Property Rapid Prototyping. *Virtual and Physical Prototyping* 6, 1:3-31.

OLIVEIRA, M. J., A. PAIO, V. RATO AND L. CARVÃO. 2012. A living System – Discursive Wall. *Proceedings 9th European Conference of Product and Process Modelling*, Reykjavik.

PAIO, A., S. ELOY, J. REIS, F. SANTOS, V. RATO and P. F. LOPES. 2011. Emerg.cities4all. Towards a sustainable and integrated urban design. *The 24th World Congress of Architecture UIA 2011*. Japan Institute of Architects: 639-643.

PUPO, R. 2009. A inserção da prototipagem e fabricação digitais no processo de projeto: um novo desafio para o ensino de arquitetura. Ph.D. dissertation, University of Campinas.

PUPO, R., G. CELANI AND J. P. DUARTE. 2009. Digital materialization for architecture: definitions and techniques. Pp. 439-441 in *Proceedings of the Sigradi 2009 conference*.

SASS, L., and M. BOTHA. 2006. Instant House: A model of design production with digital fabrication. *International Journal of Architectural Computing*, vol. 4, n. 4: 109–123.

SHEIL, B. 2012. Manufacturing Bespoke Architecture (Paper presented at Symposium Digital Fabrication: a State of the Art, 9-10 September 2011). *Nexus Network Journal* 14, 3.

SHELDEN, D. 2002. Digital surface representation and the constructibility of Gehry's architecture. Ph.D. dissertation, Massachussetts Institute of Technology.

SOUSA, J.P. 2010. From Digital to Material: Rethinking Cork in Architecture through the use CAD/CAM Technologies. Ph.D. dissertation, UTL – Instituto Superior Técnico, Lisbon.

SOUSA, J. P., 2012. Material Customization: Digital Fabrication workshop at ISCTE-IUL (Paper presented at Symposium Digital Fabrication: a State of the Art, 9-10 September 2011). *Nexus Network Journal* 14, 3.

STINY, G. 1980. Introduction to shape and shape grammars. *Environment and Planning B: Planning and Design* 7: 343-351.

VINCENT, C. and E. NARDELLI. 2010. Parametric in Mass Customization. *Proceedings of* Sigradi 2010, 236-239.

VITRUVIUS. 1960. *The Ten Books on Architecture*. Morris Hicky Morgan, trans. New York: Dover.

WEBB, P. and M PINNER. 2011. Terra therma. Pp. 94-97 in *Fabricate: Making Digital Architecture*, R. Glynn and B. Sheil, eds. Toronto: Riverside Architectural Press.

About the authors

Alexandra Paio holds a degree in Architecture (U Lusíada, 1993), a Master in Urban Design (ISCTE-IUL, 2002) and a Ph.D. in Architecture and Urbanism (ISCTE-IUL, 2011). Currently, she is the coordinator of the Vitruvius FabLab-ISCTE-IUL and co-director of the course of advanced studies in digital architecture (ISCTE-IUL/FAUP). Her research interest includes shape grammar and digital fabrication. She has published articles in national and international journals and conference proceedings related to these areas.

Sara Eloy holds a B.Arch. in Architecture (1998) from the Technical University of Lisbon, Faculdade de Arquitetura, and a Ph.D. in Architecture (2012) from the Technical University of Lisbon, Instituto Superior Técnico. Her thesis was entitled "A transformation grammar-based methodology for housing rehabilitation: meeting contemporary functional and ICT requirements". Sara is a Lecturer in the Department of Architecture and Urbanism at ISCTE-IUL in Lisbon, where she teaches courses on Building Technology, CAD and Computation. Her main research interests include shape grammar, space syntax, digital fabrication, ICT, home automation, rehabilitation, building technologies, and building accessibility.

Vasco Rato is assistant professor and director of the Department of Architecture and Urbanism, ISCTE–University Institute of Lisbon. With a graduation in architecture, a M.Sc. in construction and a Ph.D. in civil engineering (rehabilitation of built heritage), he works in the field of materials, energy performance and architectural heritage. His current interests concern digital and fabrication tools to develop energy-efficient passive systems.

Ricardo Resende holds a degree in civil engineering (2000), a Master's degree in structural engineering (2003) from Lisbon Technical University, and a Ph.D. in civil engineering (2010) from the Porto University Faculty of Engineering. His master's thesis, "A Model for the Study of Rupture Scenarios Foundation Rocky Dam", was awarded the Portuguese Geotechnical Society Masters theses award for to the biennium 2003-2004. He was Research Fellow in Portugal's State Laboratory for Civil Engineering (LNEC) from 2000 to 2010. His main area of activity is rock mechanics and he has directed and collaborated on laboratory tests and in situ rock mass characterization and numerical modelling studies of superficial and underground works. His main areas of expertise include numerical analysis of three-dimensional structures and advanced constitutive modelling of materials. He has taught structural engineering at ISCTE-IUL's Master's of Science program in Architecture since 2003, having joined full time in 2011.

Maria João de Oliveira holds a Master's degree in Architecture and Urbanism (ISCTE-IUL, 2008). Since 2007, she has collaborated with Extrastudio Office. In 2010, she joined the research team of ISCTE-IUL, in partnership with the IHRU, developing the Evaluation Study and Diagnosis in the Lóios and Condado neighborhoods, in Chelas. In 2011, she became a member of the research team of Vitruvius FabLab ISCTE-IUL.

Kevin R. Klinger

Institute for Digital Fabrication
Ball State University
2000 University Ave.
Muncie, IN 47304
krklinger@bsu.edu

Keywords: design-through-
production, humanism,
collaborative design. digital
fabrication, immersive learning,
emerging media

Research

Design-Through-Production Formulations

Abstract. Ever since the topic of design and digital fabrication in architecture surfaced ten or so years ago (encouraged by organizations of ACADIA, ECAADE, SIGraDi, and CAADRIA), it has thrived as a productive strategy for advancing the discipline. Clearly, new maps have been charted in architectural discourse that will steer us toward a promising future. Beyond just developing skills to serve new methods of design-through-production, we must now question what ends this methodology serves. This writing is an attempt to chart three potential trajectories inherent in a design-through-production methodology 1) outlining an ethic of production at regional levels in light of the ocean of global information access, 2) investigating the formulation of form inherent in digital design methods, and 3) finding humanist aims through a technological lens. Finally, several pedagocical cases are offered as incremental examples to a collective body of work which applies the design-through-production methodology.

Introduction

Over the decade of the aughts, architectural discourse has charted a new course. In the wake of the digital affect on mainstream architectural thinking, we find ourselves in a great age of exploration. Research in digital fabrication has moved from the general to the specific, while simultaneously contributing to emerging discourses in areas such as manufacturing, social impact, sustainable practices, biological structures, etc. Specific work on building component design, coupled with a performance-based pragmatic rigor about durability, strength, performance, and production have provided concrete examples of *design-through-production* investigations, and led to further clarity that the "state of the art" is indeed flourishing within and without architecture.

Open methodology

This moment in the world has engendered a shift in production (physical and cultural) as significant as the Industrial Revolution, and quite possibly even the shift from the late Medieval period to the Renaissance. Although patronage such as that of the Medici during the Renaissance is not clear today, the shift in production is being led by the technological and relational capacity of information; as the work synchronizes to the latest software, it coordinates a massive exchange of information across all cultural and political territories. As such, traditional, highly specialized disciplinary boundaries and methodologies are proving inadequate to engage the complex and interconnected questions of a global society affected by information technology.

Given global connectivity, digital communication has become central in a design-through-production process. Critical team-based machine, computation, and *most* specifically, human interface strategies are essential to solve today's complex problems, which span unsystematically over and under our established methods of problem solving. William Mitchell summed up the necessity for a new methodology in how we take on contemporary design questions quite clearly:

Nexus Netw J 14 (2012) 431–440   NEXUS NETW J – VOL.14, No. 3, 2012   **431**
DOI 10.1007/s00004-012-0128-2; *published online* 27 September 2012
© 2012 Kim Williams Books, Turin

One of the huge problems with design has been the way that the lines get broken up into these traditionally defined disciplines… The big important design issues just don't fall in these categories anymore. They sprawl in messy ways across them. We [at MIT] have architects, urban designers, economists, mechanical engineers, electrical geeks, and we put them together into an intense multidisciplinary design environment… it's everyone's responsibility to contribute to everything and educate the rest of the group as necessary on the issues that you know the most about [Makovsky 2010: 52].

Connect globally / Make locally: an ethic for design-through-production

Design-through-production projects may find unique solutions in any region by customizing open and globally shared methodologies to particular local conditions; researchers connect to the global stream of information about digital design techniques, and engage local industry partnerships invited to add specific value to the feedback loop. Bringing industry partners into the collaborative early adds tangible and practical value to the design process from the outset, and provides a demonstrable service to the region by affecting industry, both in methodology and in production. No matter the region, there are established material, cultural, and industrial processes that may inform and be informed by the digital information exchange in a design-through-production process.

Form is informed by performance

To be certain, each new design-through-production project explores unique territory and contributes to the knowledge map by adding to a matrix of possible applications. In most cases, *Form* is in*form*ed by per*form*ance!—this is the great customizable mantra for this new era in design. Yet, the principles that govern the human decision-making, in light of this new kind of digitally generated work, have yet to be clearly articulated. Nonetheless, techniques and methods have expanded to create new opportunities for making architecture and related design projects. In fact, research has tended to be less about framing new principles for making digital architecture and more about adding specific cases to the knowledge base, as each new project helps to define a collective body. Alongside slowly emerging theoretical rudders (steered towards advancing emerging processes of digital design and fabrication) are the essential pragmatic applications of *production* knowledge that truly keeps this discourse afloat—locally crafted material must perform according to the laws of physics, the demands of budget, the tolerances of equipment, time constraints, local restrictions, etc. Additionally, today's digital *design collaborative* needs to be simultaneously well-versed in the interrelationships between geometry, digital modeling, parametric organization, performance simulation, and the like.

Technology and humanism

Are we experiencing a Neo-Renaissance? Indeed, there is a re-birth of sorts underway brought on by flows of information, leading to a collapse of old and constricted paradigms. Yet, the term "re-birth" is less prescient than the term "re-scripting," which suggests an undoing or reworking of traditional methods; perhaps, then, we are in the first years of a DE-naissance—an un-birth of tired methodologies, practices, and disciplines!!!

As methods for design and production are re-scripted, what DO we value? Technology is certainly the key driver of a new paradigm, and certainly, the technologist is critical to the conversation today. Along these lines (and sixty years ago), Mies Van der Rohe argued for the total interconnection between technology and architecture:

Architecture depends upon its time. It is the crystallization of its inner structure, the slow unfolding of its form. That is the reason why technology and architecture are so closely related. Our real hope is that they will grow together, that some day the one will be the expression of the other. Only then will we have an architecture worthy of its name: architecture as a true symbol of our time [Van der Rohe 1950].

Lewis Mumford countered Mies van der Rohe's position by positing that,

progress, in an organic sense, should be cumulative, and though a certain amount of rubbish-clearing is always necessary we lose a part of the gain offered by a new invention if we automatically discard all the still valuable inventions that preceded it [Mumford 1964].

The tension between van der Rohe's and Mumford's arguments presented here, is exactly the debate that we must engage today between technological and humanist approaches.

Now, technology is discussed most frequently as "informational technology," and we are seeing its influences across the broad spectrum of society. However, projects that advance only technology for the sake of technology seem to fall flat. In light of the influence of information technology on society, can we develop a humanist position to help ground solutions to contemporary problems? As we restructure how we design for complex contemporary conditions, regardless of technology, the human is always in the middle of the equation. Scientific formulations only establish the fields and conditions of operation, but a dataset without practical human application will ring hollow.

Even though scientists have the most to do with technological imagination as invention, they generally feel very little responsibility to participate in the technological imagination as part of a larger social imagination [White 2003: 121].

Steve Jobs of Apple Computers underscored this necessity for the human relevance of technology in his Apple Keynote on March 2, 2011: "It is technology married with liberal arts, married with the humanities, that yields us the result that makes our hearts sing" [Jobs 2011]. To this end, the real potential of design-through-production lies not in the *design* (input), nor the *production* (output), but rather in the human decision-making influencing a collective human impact (*through*put).

The Institute for Digital Fabrication

At Ball State University, we have a commitment to two pedagogical strategies: *immersive learning* and *emerging media*—in other words: engaging industry directly, and cultivating skills with information technology. *Immersive learning* aims to intersect classroom activity with real world partnerships. *Emerging media*, explores the latest technology in order to prepare students for our information-driven world. The Institute for Digital Fabrication (IDF) at Ball Sate University believes that these two areas add a strategy to the pedagogical formula that is critical for making a regional impact while still adding valuable knowledge within the global exchange of ideas. The Institute for Digital Fabrication encourages a digital design-through-production methodology and curriculum through the lens of multiple course-based projects. These courses are a central component of our *immersive learning* pedagogy, simultaneously preparing students to develop a skill set with digital information, while directly collaborating and sharing skills with industry partners. More importantly, this pedagogical method always promotes team-based design groups over sole authorship as a more realistic preparation for

students. Information exchange is central to the realization of team objectives. As such, communication and information sharing skills keep the collaborative productive as projects evolve by effectively managing feedback and the global exchange of ideas between local production and design. Each design-through-production problem is seen as contributing a case to the emerging discourse, whether it be about different materials, performance properties, and mostly practices of industry in regards to the material. In this way, solutions for each parameter that constitute the design problem are useful as a reference to a collective catalogue for emerging design-through-production-based practices. This particular case-based scholarship, by dealing with unique input variables, makes relevant each operative strategy of folding, carving, bending, and the like. The key is the open sharing of information, in order to affect innovation for those who are chasing similar problems. Each project carries with it a wealth of design and production formations informed by material limitations, formulating details, full-scale production considerations, fabrication devices, different form generation techniques, etc.

Collectively these projects interrogate technology, and the real hope—human considerations, that will make significant additions to the map for the future of architecture in this great new age.

Cases for the Institute for Digital Fabrication

The following projects are to be seen as case contributions to the discourse on digital fabrication using design-through-production methodologies at Ball State University in the Department of Architecture.

Case in design-through-production : Center for Media Design
Institute for Digital Fabrication

Fig. 1. Digital office customized interior

This commission used design-through-production strategies for the Center for Media Design unit at Ball State University and was overseen by Joshua Vermillion, the Operations Manager of the Institute for Digital Fabrication (fig. 1). Each station was designed primarily in Rhino and Grasshopper, by cutting small undulating strips from full plywood sheets and, by deploying a zero-waste strategy, maximizing the yield of each panel. The components were custom-machined plywood and aluminum work pods derived strategically from clients' organizational and digital workflow.

The form integrates function with dynamic gradient effects of thinness/thickness and porous/solid effects. Light becomes a secondary "material" participant by engaging the user as it appears and then disappears through changes in vantage point, and temporal shifts.

Case in design-through-production: Transformer
Institute for Digital Fabrication

Fig.2. Sensor reactive light screen

Transformer is a layered, light-responsive shading lattice, designed and fabricated by an interdisciplinary team of students and faculty (fig. 2). The system acts as an active shading device, potentially for use within a building envelope. The prototyped system is comprised of quad-shaped, polystyrene petals arrayed in overlapping, radial clusters. The petals are situated within a lightweight but rigid and patterned support armature formed from planar polycarbonate and acrylic components while housing matrices of sensors and motors. Local variations in petal movements and aperture size are controlled by simple microcontroller arrays and varying script configurations that interpret light data from photoelectric sensors to drive small servo motors. The axial rotations from the servo motors are converted into opening and closing motions for the petals via custom gears and axles.

Case in design-through-production: tetraMIN
Institute for Digital Fabrication

The hanging screen aggregate named tetraMIN consists of componentry generated from tetrahedron geometry via Rhino's Grasshopper parametric modeling plug-in (fig. 3). Comprised of laser cut polytetrafluoroethylene (PTFE) scraps, each component forms a periodic minimal surface, and is propagated into a regular pattern by a series of reflecting/mirroring operations. The PTFE is the former roof material from the RCA Dome, a large pneumatic roof stadium prior to demolition in 2008, and was donated to the studio by People for Urban Progress, located in Indianapolis, Indiana. System prototyping and fabrication was accomplished in one week by a nine-member team of Ball State University architecture students. Working with the Institute for Digital Fabrication's faculty and equipment (primarily laser cutters), the team fabricated each component with a tab-and-slot connector system to enable the assembly of the screen. Inexpensive zip ties are also strategically deployed throughout the screen assembly for structural stiffening.

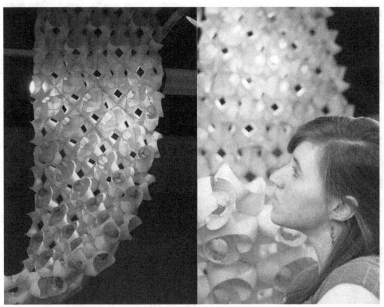

Fig. 3. "tetraMIN", PTFE woven fabric installation

Case in design-through-production: Titanium Bridge
Institute for Digital Fabrication

A digital design and fabrication seminar took on the challenge to develop entries for the Titanium Pedestrian Bridge Competition entitled "Design the Future" (fig. 4) Our interest in this program was two-fold: it was a real project located in the American Midwest (Akron, Ohio), and it was sponsored by the Defense Metals Technology Center, whose agenda was to find more civilian uses for titanium technology. The competition brief states: "Where better to find skilled competitors than from Civil Engineering, Architecture, and Industrial Design departments and schools of qualified universities in the Metals Heartland of America…" This kind of informed client, certainly contributes to making the case for building bridges towards a midwestern regional identity.

Fig. 4. Titanium Bridge: Ball State Scheme using 100% titanium

Case in design-through-production: Indianapolis 500 Hall of Fame Museum Institute for Digital Fabrication

Fig. 5. Indianapolis 500 proposal, content, and team: speed + performance

Students aligned in the first half of the semester into "innovation garages," (led by Professor Mahesh Daas), which involved collaborative teams aimed at brainstorming innovation. The second half of the semester they organized in partnership with the Indianapolis Motor Speedway and proposed an addition to the outdated Indianapolis 500 Hall of Fame Museum (fig. 5). The project goal was to deploy innovative design methodologies and fabrication techniques for a cultural institution that resonates so deeply with the Midwest's culture for production: Indiana is the home of innovation in

the automotive industry. In 1852 in Kokomo, Indiana, Elwood Haynes invented one of the first successful gasoline-powered automobiles. For a hundred years, the "Indy 500" has been the international proving ground for innovations in automobile technology since the invention of the motor-car.

Case in design-through-production: ReBarn
Institute for Digital Fabrication

A digital design and fabrication seminar partnered with regional metal fabrication experts at Zahner Architectural Metals in Kansas City, and developed a strategy to repurpose barn siding (275 unique pieces for a total of 300,000 board feet) from a one-hundred year old "Pennsylvania style" barn located near Muncie, Indiana (fig. 6). The project, developed in partnership with the mayor and local parks commission, enhances a public park along the White River. Students and Zahner discussed this project very early in the design-through-production process in order to effectively design, engineer and fabricate reBarn. This collaboration included exchanging information online, and a meeting at Zahner's office in Kansas City, and led to five water-jet cut aluminum surface panels and over 350 variable aluminum joints. Each reclaimed wood component was custom milled using a 3-axis CNC router. The digital design and fabrication technologies along with industry partnerships were instrumental in realizing the project.

Fig. 6. ReBarn: collaboration, assembly, and occupation (Zahner Metals tooled aluminum and barn siding)

Conclusions

While no clear "Medici" is directly behind the evolution of the present condition, certainly the academy plays a significant role by encouraging new skill sets and training in preparation for a globally-connected/locally-affected world. Nonetheless, many of our

present models for training and engagement are outdated. Particular to the Midwest region of the United States,

> Longworth advocates the development of a regional identity and midwestern think-tanks that will generate new ideas and a focus on issues common to the region. He also calls for a wholesale renovation of education and training... [Carr 2009: 14].

The design-through-production approach to projects can be applied to any region working with particular local conditions and sharing knowledge globally. The above *immersive learning* projects rely heavily on interdisciplinary, applied design and fabrication research, and the evolution of expertise with state-of-the-art software and devices using simulation, analysis, fabrication, and a rigorous examination of the craft inherent in digital design and production. Students connect to the global stream of information about digital design techniques, and work with consultants invited to add specific value to the feedback loop. Industry partners from the local manufacturing sector are integral participants around the virtual table as students formulate their strategies. In the design-through-production methodology, information about final production constraints is essential to initial design approaches. As such, bringing industry partners into the collaborative early adds tangible and practical value to the design process from the outset, and makes a demonstrable contribution to affecting regional industry and labor, both in methodology and in production strategies.

References

CARR, P. and M. KEFALAS. 2009. *Hollowing Out the Middle: The Rural Brain Drain and What it Means for America*. Boston: Beacon Press.

LONGWORTH, R. 2008. *Caught in the Middle: America's Heartland in the Age of Globalism*. New York: Bloomsbury.

MAKOVSKY, P. 2010. A Complete Rethink: William Mitchell and the MIT Media Lab take on one of urban America's hidden foes: the car. *Metropolis*, March 2010: 48-52.

MUMFORD, Lewis. 1964. *The Highway and the City*. New York: New American Library.

JOBS, Steve. 2011. Keynote, "Apple Special Event, March 2, 2011". http://events.apple.com.edgesuite.net/1103pijanbdvaaj/event/index.html. Last accessed 17.07.2012.

VAN DER ROHE, Mies. 1950. Technology and Architecture: a speech to IIT. In *Mies van der Rohe*, Philip Johnson, ed. New York: Museum of Modern Art, 1953.

WHITE, C. 2003. *The Middle Mind: Why Americans Don't Think for Themselves*. Harper Collins, New York, NY

About the author

Kevin R. Klinger is Director of the Institute for Digital Fabrication at Ball State University, Associate Professor of Architecture, and Director of the Post-Professional Master of Architecture program in the College of Architecture and Planning. Kevin was responsible for developing a Certificate for Digital Design and Fabrication for Ball State University. He has also served as a two-term President (03-05) of the Association of Computer Aided Design in Architecture (ACADIA), an international organization devoted to studying the advances in architecture resulting from influences of digital technology. He was an author and co-editor of the book by Routledge entitled *Manufacturing Material Effects: Rethinking Design and Making in Architecture*, developed in collaboration with Branko Kolarevic. The book assembles leading thinkers, designers, and makers from around the world to discuss experimental processes of material exploration through examining various levels of engagement and new forms of architectural production that bring designers deeper into the complexities of making, assembly, and material formulation. In coursework and through partnerships with the institute, Professor Klinger encourages explorations

in digitally-driven design, deploying techniques of digital fabrication. The Institute for Digital Fabrication operates with an ethic to "connect globally, and make locally," as it strives to both contribute to the discourse on the impact of the digital technological shift, and play a role in identifying opportunities to engage local industry and community partners. The digital exchange of information is central to this innovative process of architectural production, and demands new forms of collaboration with industry for the future of the discipline, where designers and makers are much more engaged in the total design-through-production process. As such, the Institute for Digital Fabrication is devoted to project-based collaborations that result from the intersection of emerging technology with students, industry, community, and research partners. For more information, see http://www.i-m-a-d-e.org.

Bob Sheil

The Bartlett School of
Architecture
University College London
22 Gordon Street
London WC1H 0QB UK
r.sheil@ucl.ac.uk

Keywords: design,
manufacturing, bespoke
architecture, digital fabrication,
sixteen makers, design through
production

Research

Manufacturing Bespoke Architecture

Abstract. At the disposal of today's architect is an evolving array of interoperable tools and processes that allow the fabrication of design propositions to be increasingly complex, non-standard and adaptive. How are we equipped to deal with such a growing breadth of new potential, and how are the philosophies that underpin this potential being defined? This paper attempts to address what is something of a contemporary dilemma in architecture, as the constraints of industrial standardisation are relaxed. Have the roles of designers and makers changed in a way that we've not experienced before, and is a new approach to making architecture emerging?

Part One

Over the past decade, conventional protocols of exchange that focused on the key relationship between design and making have been thoroughly redefined by digital technology [Sheil 2005a]. For centuries, the construction of prototypes, artefacts, buildings and structures has operated on a rolling tradition of visual and verbal communication between designers, consultants, makers, clients, users, regulatory bodies and contractors. In making buildings, roles were defined by where individuals and disciplines were located on a chain from concept to execution. All were reliant on its links being successfully forged, not only to achieve results, but also to underpin their status within their respective professions and trades. Prevailing over the entire process was the design, an assemblage of cross-referenced visualisations, specifications and quantities forming the templates and instructions for making. Given the numeracy of complex transfers from one step to the next, constructs in architecture have evolved as negotiated translations; the most engaging are those that have recognised this in a creative and informed way from the very outset.

The redefinition of these historic protocols was initially led by the gradual adoption of computer-aided drawing in the early 1990s by practice and academia. As three dimensional modelling and rendering became more available and sophisticated, a frenzy of liberated experimentation ensued. Speculative design looked to the weightless and scaleless domain of digital space as the new terrain for innovation and speculative discourse and as the means to compositionally define spatial and formal complexity.[1] The gap between the designer's vision and operations of the construction industry widened as fabrication processes remained largely analogue in how they were driven and delivered. A defining example of this challenge was Future Systems' Media Centre at Lord's Cricket Ground (competition winner 1995, opened 1999), in which the primary enclosure was entirely prefabricated by the Pendennis Shipyard in Falmouth, Cornwall, as theirs was the only industry both familiar and experienced in extrapolating design information for the fabrication of such forms.

Concurrently though, news tools of computation, means to capture and analyse the performance of buildings, built environments and the behaviour of users, brought a fresh understanding of the complexity and density of dynamic contexts in architecture.

Geometry was re-ignited as a great organiser, only now it was adaptable and smart, as developments in design software far outstripped those in the world of how such forms could easily be made; more significantly, the means to communicate from one realm to the other was restricted. In the first decade of the new millennium this restriction started to lift as CAD/CAM (computer-aided design and computer-aided manufacture) entered the mainstream. Subsequently, a vast expansion on the remit, scope and potential of the designer was released, allowing for their direct engagement and control of fabrication processes. Likewise, the capability of manufacturing and construction to fulfil design intent was expanded, and a creative dialogue between design and fabrication began converging once more.

Size Matters

The term 'bespoke' is said to have originated over one thousand years ago from the old English 'bespeak', meaning 'to request', 'to order in advance' or 'to give order for it to be made'.[2] Tailors of London's Savile Row claim that the term was in common use on their street from the seventeenth century, when tailors kept their cloth on the premises and customers would 'bespeak" a particular length of fabric to be fitted as a suit or uniform. The first recorded use is thought to have occurred in C. Clarke's *A Narrative of the Life of Mrs Charlotte Clarke* (London, 1755) on the life and experiences of an actress, cross-dresser and famous playwright's daughter in eighteenth-century London. In this instance, the term was used in reference to the performance of a 'bespoke play', in the sense that it was a one-off.

Almost 275 years after suits and uniforms were first made there, the Savile Row Bespoke Association was founded in 2004, and within three years they trademarked the term 'Savile Row Bespoke' which defined a two-piece bespoke suit as 'crafted from a choice of at least 2,000 fabrics, be made almost completely by hand, and requiring at least 50 hours of hand-stitching'. To qualify, Savile Row Bespoke suits must also be 'derived from a paper pattern, individually cut and produced by a master cutter, and subsequently undergo personal supervision by the master cutter in the course of production'. In June 2008, the association lodged a complaint under the truthfulness rule at the UK's Advertising Standard's Authority (ASA) against the international firm Sartoriani who had recently opened a nearby store where machine cut suits were promoted as bespoke.[3] The ASA noted the complainants' argument that 'the advertised suits were machine-cut abroad[(in Germany at the time] to a standard pattern after initial measurements were taken and adjusted at the end of the process' and that the suits 'at best' should be described as 'made to measure'. Sartoriani claimed that the initial machine-cut fabric pattern was a 'working frame' that could be individually adjusted if the customer's measurements did not match a standard pattern size, and that this occurred in some cases.

The ASA concluded that, following recent changes to the industry, the use of the word bespoke to describe the advertised suits was 'unlikely to mislead'. They went on to say, 'both bespoke and made-to-measure suits were "made to order"', in that they were made to the customer's precise measurements and specifications, unlike off-the-peg suits". The ASA did not rule on a fuzzy distinction between hand-made or machine made, nor the particular differences of approach either method adopts in making or fitting, nor even where the suit was made; they ruled on the rather neater and universal principle of measurement.[4] Whether the artefact is made to order, made on the premises,

made to measure, made by hand, or manufactured by machine, our recent understanding of the centuries old term 'bespoke' has undoubtedly been altered.[5]

Designer as maker

Today, the bespoke is referred to in the context of personalised stationary, customised software, pharmaceuticals, wines, cars, financial investments, even biscuits, suggesting it is increasingly common for everyday and mass produced artefacts to be made to order. Bespoke is also a term associated with architecture, and in the first instance through the idea that most buildings are in some sense unique to their location, the time they were designed and built, who they were designed and built by, how they were built, and the circumstances that surround their occupation and use. More specifically, the architecturally bespoke has associations with craft, ornamentation, materiality, fit, uniqueness and the unrepeated. However, as with the core productions of Savile Row, and for many of the same reasons, the meaning of bespoke in architecture in recent years has shifted on methodological grounds. Key to this are two primary issues: first, radical changes in how architecture is designed and made; second, a vast expansion in what might be regarded as materials for architectural specification – from the deployment of nanotechnologies to choreographing four-dimensional behaviour.

With regard to the background of the former, one of the key properties of the bespoke is the involvement of the maker as the designer. Unlike tailors, most architects do not make the things they design; they make design information, the equivalent of making the pattern, not the suit. Architects, however, also make space, an immaterial substance, and the equivalent of forming the host upon which the pattern is draped. The architecturally bespoke is therefore associated not only with the ability to establish rules for the 'made to order', but with the generation of a design that understands and anticipates the challenge and consequences of making that particular order, and its role within a construct that is greater than the sum of it parts. Equally, the subsequent bespoke artefact reveals the manner in which this commission has prompted a skilled craftsman or specialist to respond, and how the overarching construct fits together. At its best, this collaborative dialogue has the capacity to transcend the drawing as a literal template and goad the maker and the material into new territories. Since the introduction of digital fabrication technologies, profound changes have been brought upon this relationship and the habitual protocols between design and making. Through the progressive elimination of craftsman and skilled machine operatives, the expertise that designers relied upon to translate their work has diminished and in many respects transgressed into their domain.

A further ingredient of the bespoke is how an intimate knowledge of materials and their performance in use informs design, and relate to its method. Cutting fabric on the bias and incorporating ease within a garment each have their equivalent at architectural scales, such as the imprint of shuttering on cast concrete, or the deflection of steel under load. In tailoring a bespoke suit, drawing and making are synthesised from the outset where graphic and illustrative representation dissolves as the final artefact appears. Processes of measurement, pattern-making, cutting, forming, and joining components progress through three stages of fitting and fabrication known as the 'Skeleton baste',[6] the 'Forward' and the 'Finish bar finish'. Such intimacy between drawing and making must therefore anticipate and understand the difference between the simulated and the actual and adapt accordingly. Clearly, the range of materials specified in architecture is substantially

broader and more complex than that of tailoring, not least as it also incorporates strategies for immaterial qualities such as context, light, reflection, temperature, sound, culture, meaning, memory and emotion. In this regard, both the tools and the palette of the twenty-first-century architectural designer are rapidly expanding as they provide the ability to approach design as a strategic act with novel outcomes.

Part Two

Two co-authored design and make projects are examined here; a chair from 1995 (figs. 1-2) and a small building from 2009 (figs. 3-12). Both projects fit within the author's definition of protoarchitecture (see [Sheil 2008]), a self-coined term to identify experimental design that challenges the methods and role of the designer particularly in relation to how and why the work is made. Secondly, and central to the arguments presented here, both projects identify a key transition in the definition of the bespoke that spans a period of significant change in design and fabrication methodologies and tooling. In analysing these projects together for the first time, it will be argued that many of the strategies in designing and making a bespoke piece of furniture that went beyond the realm of the conventional drawing, and were exclusively developed by hand, are now adoptable through digital design and fabrication technologies. It will also be suggested that these new facilities must be seen as essentially hybrid disciplines that are practiced adjacent to the point of production. What is also being explored here is an underlying idea that integrated digital design and fabrication technologies have instigated a renewed relationship between the bespoke and the prototype, and that exploration of either presents opportunities for the other. What is new in this relationship is that these pursuits can be exercised mutually and synthetically; for those who wish to take advantage of this potential, there are significant implications for the way they might practice and learn [Callicott 2000].

Making the Bespoke

The chair shown in figs. 1 and 2 was sixteen*(makers)[7] first commission and was made for a management consultant in 1995 who noticed the collaborative's experimental approach to design and making on projects 'plot 22' and 'Dartmoor' [Groat 1996]. The practice was established whilst Callicott and Sheil were students at The Bartlett School of Architecture UCL. Both were midway through their undergraduate studies when Peter Cook was appointed as the school's Professor of Architecture in 1990; under his charismatic stewardship the school became a vivacious and inventive forum for experimental ideas. The 'fee' for the chair was a 50% barter on our client's second-hand Macintosh Powerbook 100 (which was introduced on 21 October 1991). One handmade chair in part exchange for a laptop; it seemed the analogue and the digital were destined to be present in the history of our practice from the start. At the time of the commission our client was practicing the Alexander Technique, a method of focusing and developing controlled body posture and balance habitually and intuitively. The commission was envisaged as a means to address this practice routinely, and to design an everyday point of support for reading, typing, dining and relaxing. The particular movements and strains upon forearm, eye, head, neck and shoulders in relation to the gizmos of the day, such as the centrally placed traceball, a mouse, a 23 cm, 600 x 400 pixel resolution backlit LCD display with compact keyboard, were absorbed as essential but not exhaustive design criteria.

Fig. 1. 'Webb Chair' (1995), sixteen*(makers)' first design and making commission, developed through a series of conversations between authors and client. © sixteen*(makers)

Fig. 2. 'Webb Chair and Webb Table' (1995), Sixteen*(makers) first and second design and making commissions, developed through a series of conversations between authors and client. © sixteen*(makers)

Somewhat in the traditions of Savile Row, the client was surveyed, his frame and posture measured and noted, and other chairs and stools were commandeered as adaptable props. They were made higher, wider, more vertical, softer or cooler. Likes and dislikes of the function and form of chairs were identified and key objectives were agreed upon and marked on jigs, rulers, floors and walls. Issues of comfort and restraint were explored in relation to the chairs purpose and role, and environments where the chair would likely be located were noted for matters such as deflection and vibration. Materials from which to make the chair were also explored, not only for performance and visual preference, but also for their capacity to carry other narratives within the object. Lying about the studio in Shoreditch at the time were a number of speculative test pieces in mild steel, hardwood, acrylic, glass and rope, forming a haphazard library of experiments and assemblages in relatively simple techniques. Amongst these was an ongoing trait to customize materials, particularly steel, from their origin as extruded standard profiles, into ends, junctions, and limbs through abrasion and forging. Some of these traits had a highly graphic as well as tactile quality, portraying not dissimilar looks to those we once pursued in ink. However, it was their tangible properties, such as weight, surface quality, conductivity, resonance and reflectivity that were the predominant objects of investigation and value.

From this catalogue of references, the designs first move was established in the form of a foot adapted from a short section of heavy rolled T-bar in mild steel. To the central web of the T-bar, a flat bar the same thickness and width as the central web was beveled and arc welded. The joint was ground, filed and sanded until seamless and the shiny surface was reheated with an oxy acetylene torch until black again. Overgrinding was always something to be mindful of, for as soon as the weld surface dipped below the adjoining sections, the only course of recovery was to break the weld and start again. The inner and outer radii of the rolled T-bar and the feathering of its tapered flanges, set a geometric tone for key positions where the chair's limbs would end or meet others. The length of this first element, about 1m, far exceeded what was required, and so it remained as the key datum from which to strike the seats horizontal axis until the remaining substructure of the chairs legs and spine was complete. Both front legs also stemmed from the same flat bar, but here they were jointed to a parallel square section of the same thickness. The heavier elements connected vertically to form the seat's substructure, whilst of the lighter, one swerved away on the left hand side to receive a rigid arm, and the other tapered to finish short of a cantilevered right hand arm with some give.

When our client returned for his 'skeleton baste' fitting, the chair's key dimensions and alignments were tested and adjusted. Movements and postures were simulated and marks were made on the chair's steel carcass in chalk. Rough boards were fashioned as jigs to form seats, and various blocks and wedges were positioned to judge the thickening of various elements such as arms and the spine. The torch was re-ignited and geometries of the frame adjusted before the client departed. As the chair passed through two more fittings, a small pile of reclaimed tonged and grooved oak strip flooring was rescued to form the seat and subsequently jointed to form two panels tied to the subframe with a leather thong. The rigid left arm of the chair was filled out with a further section of jointed and reclaimed oak and the flexible right arm was wrapped in leather from a local upholstery merchant. Our studio was located at the western end of Sunbury Workshops in London's historical quarter of Shoreditch. In the two units beneath us were a shopfitter and an octogenarian wood turner. Next door below were silversmiths working above a frame gilder, and further along a prestigious upholsterer, and a team of

glassblowers. Although,not formally organized, the range of skills being practiced was highly influential and complimentary to our aims as architectural graduates. We could witness at close proximity the vibrancy and vulnerability of an urban micro-industry, and realise how little our education did to make us aware of how direct engagement with its potential could inform our design strategies. As the project of a fledgling experimental design practice, these acts of meticulous production informed many of the foundations of our attitude to repositioning the designer as maker. However, it was clear that in terms of a critical place in our chosen industry, they resembled more the world of the nineteenth-century artisan than the contemporary professional.

Bridging the gap

Resolving the bridge to this divide would take us back to our exchange fee for the chair and the presence of a new realm for workshop tooling, that of the digital. As an indication of how fast changes in related technologies took hold in this time, Nick Callicott's *Computer Aided Manufacturing in Architecture: The Pursuit of Novelty* [2000] was published just five years and numerous operating system and hardware upgrades later. The subsequent portfolio of sixteen*(makers) increasingly addressed the evolving relationship between digital design and making through a series of speculative projects and publications, including 'Cut and Fold', 'STAC', and 'Blusher' (all 2000-01), *Design through Making* [Sheil 2005b] and *Protoarchitecture* [Sheil 2008]. In parallel, their work shifted further towards a directed stream of academic research on both digital and analogue practices, with Ayres and Leung embarking on related doctoral studies with particular focus on responsive systems, persistent modelling [Ayers 2012], and environmental analyses. Greatly facilitating further opportunities for academic collaboration with industry, Callicott left UCL in 2005 to set up Stahlbogen GmbH in Blankenburg Germany, with partner Kris Ehlert. At the time, sixteen*(makers) were establishing speculative proposals as architects in residences at Kielder Water and Forest Park Northumbria, on behalf of The Kielder Partnership. With an open brief, the appointment provided an extended period of speculative investigation on the difference between digital modelling (as the 'ideal') and physical installation (of the 'real'). On the basis of a series of built installations on remote sites, the work, which led to a solo exhibition at The Building Centre in 2007 entitled 'Assembling Adaptations', presented the potential for real-time monitoring of microenvironmental change to drive the design of a bespoke dynamic architecture [Sheil and Leung 2005]. Whilst this research remained ongoing and formed the core of both Ayres's and Leung's doctoral research, the practice was approached by The Forestry Commission to design a public shelter on a proposed perimeter walkway to the reservoir. The additional brief carried a set of performance requirements and budget that would restrict many of the immediate possibilities of the residency. However it offered the first opportunity to collaborate with Stahlbogen GmbH from the outset of a new project, and a fresh set of investigations were embarked upon.

Manufacturing the bespoke

Completed in June 2009, the shelter known as 55/02 is named after its coordinates: 55° 11.30 N, 02° 29.23 W, otherwise known as Cock Stoor, Lakeside Way, Kielder Water and Forest Park, Northumberland, UK.[8] Key issues to recall from theses accounts relate to:

1. winning the commission on the basis of strategic information rather than an illustrated design visualisation, to clearly posit an underlying intent to develop the built design on highly specific qualities of the final and 'real' site;

2. making it clear in the award of the commission that the built design would explicitly evolve in collaboration with Stahlbogen GmbH.

Qualities of the selected site that informed the primary aims of the work included the intersection of distant harvest lines, trenches, gouges and trails, and the atypical qualities of the site that overlooked the reservoir from a raised mound on the north shore. They also included orientation to prevailing weathers and sunlight, relationship to key views and distances from other resting points, and the particular spatial qualities of the surrounding plantation of preserved Scotts pine and their array of geometric intersections through the loose grid of straight and bare tree trunks, each between 5° and 15° off plumb. These, and the rich matrix of adjacent layers and nodes of orientation, generated a dynamic field of variable spatial depth and quality of light, sound, temperature and enclosure for the design to acknowledge. Collaboration with Stahlbogen GmbH was initiated from the very outset of the commission, with the manufacturers in attendance and taking part in the site analysis and selection visit. In this respect, access and manoeuvrability of plant and personnel was also considered, and it was thereby possible to renegotiate the proposed route of the Lakeside Way and determine the position of a planned temporary road spur for the supply of materials to build the new pathway. At the conclusion of the site visit it was agreed that the shelter should have no obvious boundary between its material edges and the rich constellation of its environment, and that in response to its exposed position it should offer substantial shelter to the north, with a greater degree of openness to the south.

In recognition of unnegotiable budget constraints, a key early decision was to ensure that the design took account of Stahlbogen's existing tooling and expertise, that the number of materials be limited, and that Stahlbogen's established knowledge in sheet steel fabrication be exploited and stretched. On this basis, the design progressed through simultaneous speculative exercises in drawing and making which flowed between the factory in Blankenburg and studios in London and Copenhagen (from where Phil Ayres collaborated). Flowing from one workstation to the next were images, drawings and comments on 1:1 physical prototypes produced by Stahlbogen, spliced with sketches, hand-made models, 3D drawings and 3D prints emerging in London and Copenhagen. This sequence led to the evolution of the projects distinctive 'tank-like' walls, the idea for which was led by a series of 1:1 test pieces at Stahlbogen. These units not only distinguished much of the project's formal character, but also informed the development of speculative design drawings and models, and established the unconventional order of exchanging design information between the drawing makers and the manufacturer. Once primary decisions on how to approach the fabrication and design of the enclosing walls was established, an entirely fresh CAD model was constructed through which every resulting form was assessed for potential conflict with the dimensions and limitations of the equipment and processes that would produce them.

Fig. 3. 'Shelter 55/02' (2009) by sixteen*(makers) in collaboration with Stahlbogen GmbH. The illustration shows the first test prototype unit in manufacture at Stahlbogen's works in Blankenburg Germany. © sixteen*(makers)

Fig. 4. 'Shelter 55/02' (2009) by sixteen*(makers) in collaboration with Stahlbogen GmbH. The illustration shows Nick Callicott inspecting the work and refining the design prior to completion at Stahlbogen's works in Blankenburg Germany. © sixteen*(makers)

Production design evolved by constant shuffling between fabrication, drawing, animation and assembly. Once information for the tank walls was fully developed, the steel was cut, folded, capped, welded, and connected as a structural unit in a few days. Assembled in the factory in the same configuration they would finally be placed on site, they became, for a period, models upon which to judge and test propositions for overhead roof elements. Strategies for these elements were visually developed upon the CAD file, and cross-referred to both 1:20 physical design models and the 1:1 partially built shelter of tank walls.

Fig. 5. 'Shelter 55/02' (2009) by sixteen*(makers) in collaboration with Stahlbogen GmbH. A detail of the roof as installed on site at Cock Stoor, Kielder Forest and Water Park, Northumberland UK. © sixteen*(makers)

Fig. 6. 'Shelter 55/02' (2009) by sixteen*(makers) in collaboration with Stahlbogen GmbH. The illustration is created from a 3D LIDAR scan captured by Scanlab Projects at UCL. © sixteen*(makers)

In this sense, the design exercise became truly synthetic with making and the methodological strategies of the projects 1995 forerunner were resurfacing, albeit now equipped with considerably more adaptable and flexible design tools. Here, design decisions were being made and becoming available that would otherwise be excluded from the designer, who might only have provided an initial set of drawings to be followed as instructions for making. In this instance, such information provided by the team was only being regarded as a prompt and a guide, and final decisions were formulated on how the design implied in those drawings might operate both visually and feasibly at 1:1 on the shop floor. Crucially, this critical analysis and subsequent decision was being taken by a designer with ownership over the design and advanced skills in production.

Subsequently the design of each of the roof elements as built, evolved substantially from initial design proposals as developed in model or CAD form. A number of factors played into this outcome, including considerations on physical and visual weight, soffit qualities, complexity and physical geometry of surface in relation to vertical elements, and the feasibility of assembly, with very limited resources on site at Kielder.[9] In addition, and perhaps one of the most surprising points to make, is how the final iteration for the design of the roof elements was informed by the visual language that was emerging from the preceding completed elements more than that which was conveyed in initial drawings. This point can best be understood by comparing the somewhat stark quality of the projects production drawings and the more fluid quality of the built works aesthetics. This is not to say that that a fluid aesthetic was found only through making, but that the production drawings that were necessary to construct the work were not being judged in this way.

Fig. 7. 'Shelter 55/02' (2009) by sixteen*(makers) in collaboration with Stahlbogen GmbH. The illustration shows the roof section in transit at Stahlbogen's works in Blankenburg Germany. © sixteen*(makers)

Figs. 8-9. 'Shelter 55/02' (2009) by sixteen*(makers) in collaboration with Stahlbogen GmbH.
Left, View of the work part complete at Stahlbogen's works in Blankenburg Germany; right, Detail
view of the work part complete at Stahlbogen's works in Blankenburg Germany
© sixteen*(makers)

Such drawings were made to provide the necessary digital information to feed the digital manufacturing process. In a sense, they partially equate to traditional shop drawings where in the past makers would redraw the designer's drawings at 1:1, and through the act of drawing the makeability of the design was established. However, in this instance, with the maker inhabiting and continuing the role of the designer, production drawings were constructed in order to facilitate a desired outcome in making, and if in view or in print the subsequent drawings were deemed to be aesthetically odd or unappealing, this was ignored in the knowledge of the results they delivered. This aspect can be understood globally in plan and section, which is a form of information on the project typically requested by publishers. Through the conventions of reading such literal projections, these interrogations struggle to capture the central qualities of the built work. The graphics that are generated greatly overstate gentle folds as harsh creases, and the resultant complexity of overstated lines and geometries are all that the eye sees. Little of the subtlety that easily transmits from the formed and assembled materials is conveyed, and it is very doubtful that had such drawings been presented to the client in the first instance, that the commission would have been awarded. By letting go of the design drawing as the primary instruction for making, and by reading such initial drawings solely as strategic intentions of spatial and material organisation, an otherwise unexplored architectural response is found in the place and process of production.

Fig. 10. 'Shelter 55/02' (2009) by sixteen*(makers) in collaboration with Stahlbogen GmbH. Installing the roof at Cock Stoor, Kielder Forest and Water Park, Northumberland UK. © sixteen*(makers)

Fig. 11. 'Shelter 55/02' (2009) by sixteen*(makers) in collaboration with Stahlbogen GmbH. Installing the roof at Cock Stoor, Kielder Forest and Water Park, Northumberland UK. © sixteen*(makers)

Fig. 12. 'Shelter 55/02' (2009) by sixteen*(makers) in collaboration with Stahlbogen GmbH. View of the work as installed at Cock Stoor, Kielder Forest and Water Park, Northumberland UK. © sixteen*(makers)

Part Three

A year to the month after 55/02 was assembled and installed on site, the building was captured as a 3D lidar scan using equipment on a loan to UCL by the international instrument and measurement manufacturers FARO.[10] This technology has been developed for precision engineering and industrial applications over the last decade, but only in recent years, and coinciding with the evolution of point cloud modelling, has it entered the domain of the design and construction sector. With the potential to provide users with rapid and accurate feedback on existing structures in the form of exportable digital models, the capacity to exploit the resource of accurate and useable measured data is broad and deep. In this instance, it provided the authors of 55/02 with an asset from which to trace and record decisions that had not been developed through drawing. The scanned model may be overlaid on the CAD model and anomalies identified. In this instance such data would not be regarded as a fault, bur rather as a difference that was not documented. One can easily imagine, however, that a less positive conclusion might be drawn in other circumstances. As in a world of increasing regulation and accountability, design differences that are undocumented tend to be labelled as 'faults' or at the very least 'disputes'. Drawing, whether digital or analogue, is without doubt an essential process and tool in the production of architecture. Both the act and the product of drawing are also central in defining the role of the designer, but as I hope this exploration has conveyed, in an age where we are bypassing the translation of drawings by those with the tacit skills in how they are realised, we might wish to consider their bearing strength in this great task, and what we can do to ensure their best intentions are fulfilled.

Acknowledgment

This essay has been adopted from the author's leading chapter in the publication *Manufacturing the Bespoke: Making and Prototyping Architecture* [Sheil 2012].

Notes

1. Seminal publications of this period that explore these themes include a number of guest-edited issues of *AD* including *Folding in Architecture* [Lynn 1993] and *Architects in Cyberspace II* [Spiller 1998].
2. See http://dictionary.reference.com/browse/bespoke?s=t&ld=1032.
3. See the ASA report on the judgement at: http://www.asa.org.uk/ASA-action/Adjudications/2008/6/Sartoriani-London/TF_ADJ_44555.aspx.
4. Recent developments in three-dimensional body scanning offer a route to broader and more perplexing questions on specific measurement [Treleaven 2004]. Here, not only is the individual accurately measured in 3D, but the data may also be of assistance in monitoring their state of health, contributing to national welfare statistics and understanding their physical attractiveness to others [Swami and Tovée 2005; Tovée et al. 2002].
5. Shortly afterwards Sartoriani went on to open a store on Savile Row; however, at the time of writing the firm had ceased trading and existing orders had been taken over by online trader asuitthatfits.com
6. Skeleton baste is a term used in tailoring to describe the first and rough fitting of a bespoke suit. The second fitting is known as 'The forward', and the final as the 'Finish Bar Finish'.
7. Sixteen*(makers) operate between practice and research, and between design and making. Established at The Bartlett School of Architecture in the mid 1990s, the group includes Phil Ayres, Nick Callicott, Chris Leung, Bob Sheil and Emmanuel Vercruysse. Throughout its life, members of the group have remained independently active in academia, practice and industry, as they have worked together on a project-by-project basis from respective bases in London, Paris, Copenhagen and Blankenburg. Callicott has since gone on to establish Stahlbogen GmbH, a subsidiary of Ehlert GmbH, in the Harz region of Germany, whilst Ayres, Leung, Sheil and Vercruysse continue their ties with the Bartlett in design, research and teaching. The group now operate as an architectural consultancy at UCL. Further details may be found at www.sixteenmakers.com.
8. For a general analysis, with background information on local context and project sequencing, see [Sheil and Leung] and [Sheil 2009].
9. The site at Cock Stoor is 4 miles from the nearest car park and delivery point, is accessible via an unstable forestry road or by boat, and is without power supply. At the time of installation, it was also beyond the reach of a mobile phone signal.
10. The scan was carried out and post-processed by Matthew Shaw and Will Trossell of Scanlab Projects.

Bibliography

AYRES, P., ed. 2012. *Persistent Modelling: Extending the Role of Architectural Representation.* London & New York: Routledge.

CALLICOTT, N. 2000. *The Pursuit of Novelty: Computer Aided Manufacturing in Architecture.* London: Architectural Press.

GROAK, S. 1996. 'Board Games', a profile of sixteen*(makers). In *Integrating Architecture*, Neil Spiller, ed. *Architectural Design* (Profile no. 123): 48-51.

LYNN, Greg, ed. 1993. *Folding in Architecture. Architectural Design* (Profile no. 102). London: Academy Group Ltd.

SHEIL, Bob, Nick CALLICOTT, Phil AYRES and Peter SHARPE, eds. 2011. *55/02 A sixteen*(makers) Project Monograph.* Toronto: Riverside Architectural Press.

SHEIL, R. 2012. *Manufacturing the Bespoke: Making and Prototyping Architecture.* London: Wiley.

SHEIL, R. 2009. A Manufactured Architecture in a Manufactured Landscape. *Architectural Research Quarterly* 13, 3-4: 200-220.

SHEIL, B., ed. 2008. *Protoarchitecture: Between the Analogue and the Digital. Architectural Design* 78, 4 (Profile 194). London: Wiley.

SHEIL, R. 2005a. Transgression from Drawing to Making. *Architectural Research Quarterly* 9, 1: 20-32, 26.

SHEIL, B., ed. 2005b. *Design through Making.* Guest issue of *Architectural Design* 75, 3 (Profile 176). London: Wiley.

SHEIL, B. and C. LEUNG. 2005. Kielder Probes – Bespoke Tools For An Indeterminate Design Process. Pp. 254-259 in *Smart Architecture*, O. Ataman, ed. ACADIA (Association for Computer Aided Design in Architecture). Savannah, GA: Savannah College of Art and Design.

SPILLER, Neil, ed. 1998. *Architects in Cyberspace II, Architectural Design* (Profile no. 118). London: Academy Group Ltd.

TOVÉE, M. J., P. Hancock, S. Mahmoodi, B. R. R. Singleton and P. L. Cornelissen. 2002. Human female attractiveness: Waveform analysis of body shape. *Proceedings of the Royal Society of London,* series B, 269: 2205-2213.

SWAMI, V., and M. J. TOVÉE. 2005. Female physical attractiveness in Britain and Malaysia: A cross-cultural study. *Body Image* 2: 115-128:

TRELEAVEN, Philip. 2004. Sizing us Up. *IEEE Spectrum* 41, 4: 28-31.

About the author

Bob Sheil is the Director of Technology and Computing at the Bartlett School of Architecture, University College London, where he leads MArch Unit 23 with Emmanuel Vercruysse and Kate Davies. He is a founding partner in sixteen*(makers), a workshop-based architectural practice, whose work has been widely published and exhibited internationally. He guest-edited AD Design through Making (July/August 2005), AD Protoarchitecture (May/June 2008), and the AD Reader Manufacturing the Bespoke (2012). He has also Co-Chaired and Co-edited the International Conference and Publication FABRICATE (2011) with Ruairi Glynn. He has also recently published a monograph on sixteen*(makers) 55/02 project through Riverside Architectural Press (2012).

Tomas Diez

Calle Pujades 102
08005 Barcelona, SPAIN
tomasdiez@iaac.net

Keywords: Fab labs, digital
fabrication, third industrial
revolution, self production,
3D printing, innovation,
new manufacturing tools,
fab lab Barcelona, urban
design, smart cities,
urbanism

Research

Personal Fabrication: Fab Labs as Platforms for Citizen-Based Innovation, from Microcontrollers to Cities

Abstract. The "digital fabrication" revolution being lived today, both in knowledge creation and in technological developments will become more than a simple formal exploration in architecture and design, or a set of tools exclusive to advanced industries. New tools and processes are becoming more accessible to the masses and are being shared all over the world through Internet platforms, with an open source philosophy, both in software and hardware. The collective mind that is being empowered everyday will define the future of production in the life of mankind and its relation with the environment. The role of architects, engineers, designers and many other professionals, will be reshaped and reconfigured to fit into new models of production and creation. These will need to be supported by new manufacturing platforms, knowledge generating and sharing know-how.

Introduction

A look back at history:

- *ca. 1400 years since Gutenberg invented the press;*
- *more than 250 years since the industrial revolution changed the way we live;*
- *more than 70 years with CNC machines;*
- *ca. 30 years with personal computers;*
- *ca. 20 years since the Internet revolution;*
- *twenty years ago we had cameras that worked with rolls. Once we took photos, we rewound these rolls and we had a development lab. These laboratories required twenty-four hours (then only one) to reveal the pictures. Do you remember the last time you saw a development lab in your town?*

Consider the following questions:

- *Which information do we produce?*
- *What is being produced in cities?*
- *Which is the balance of production-consumption of goods and energy in our every day life?*

History shows that tools have been used for over 2.6 million years; they are extensions of our physical capabilities and at the same time are the interface with our natural environment or habitat. The earlier tools helped humans get provisions of food by hunting and gathering. This in turn was converted into energy needed to perform other activities and to evolve both body and mind over time. Hunting tools made human evolution possible and are one of the milestones in our history. The development of tools provided additional capabilities in terms of the relationship with the environment; they helped in protecting against all types of weather, establishing permanent sources of food

DOI 10.1007/s00004-012-0131-7; *published online* 5 October 2012

(agriculture), and simply improving the quality of life for early mankind. Humans established fixed settlements where constructed structures protected the inhabitants from the weather and predators; this allowed for the development of the first complex social structures, built on some phenomena still familiar today: clan or community, trade or market, specialization, diversification, etc. Today's society is based on a very basic fact: mankind was able to gather around specific resources, both natural and man-made. The first human settlements became the platforms for the exchange of knowledge and goods.

Technology has always existed, in the past this was referred to as tools. A shell used to produce a specific sound for long range communication between aborigines in Oceania is not that different from a WiFi enabled device that allows New Yorkers to create blogs. Both shaped (and still shape) landscapes: the landscape created by social interactions that are not bound solely to physical presence, and the physical space needed for social interaction. We shape tools but tools shape us as well, not always in the way we envisioned whilst designing them, or not for the purpose they were created to serve. A sharp rock tied to a stick was used to hunt animals to provide food and energy, one way this tool has evolved is into high-tech knifes used by sushi masters to cut fish; however, this is definitely not the same knife used to "hunt" the fish or to capture it. A chair was not a "chair" until it was so named; if a designer creates a chair (one of the objects used most often in everyday life) it is very likely not being designed to be effective as a weapon against enemies in a bar fight (with some luck these will only be seen in the movies). In the same way, when a journalist's shoe (maybe designed by Armani©) is used as a flying object aimed at hitting a US president's face (who reacted faster than Bruce Lee to avoid being hit) it is assumed that this was not the initial purpose the shoe was designed for. These examples aim only illustrate the idea that a tool evolves, mutates and changes its use in relation with the user, and there is a strong backward and forward relationship during this process.

On a broader scale than a shoe or a chair, the evolution of tools has allowed humans to create more complex and advanced construction systems and therefore develop infrastructures to enable more solid and longer lasting human settlements. An efficient and precise control of materials (from wood to minerals, stone, etc.) and techniques (structural development, on-site construction methods) brought a new scale to construction; it was the beginning of civilization. The Romans did not conquer the world; they civilized it. The Roman Empire would not have existed as we know it without its monumental construction scale. Water infrastructures, roads, coliseums and other mayor buildings shaped a physical scenario for power without limits, driving society towards the current Western civilization model, and having effects which echoed in the rest of the world. Maybe this was the beginning of globalization. If we consider that America was colonized under Roman models – Cardus and Decumanus were the starting point for every new city created in the sixteenth and seventeenth centuries, and that these are still today's city centers –, and that the Middle East and Asia, have been adopting a Western developmental model, then we have an (almost) homogenized city model in a large number of important cities around the world.

The construction of ancient Roman cities depended on decisions made in Rome. The centralized power defined the strategies and models for the expansion of the empire: how, when and where a city would be invaded, created or occupied. Inhabitants were traders, artisans, entertainers, and were represented in the Senate, their voices resounded at the Coliseum. If one were to talk about religion a similar analogy could be made; however, this is outside the scope of this essay. Centralized power produced cities during the last (at least) two thousand years; that power was expressed within the cities in the name of

the gods, religion and, more recently, capital and the market. Decision-making and knowledge ownership resided in protected libraries or was based on a cost-effective model that granted access only to those who could afford it. These models seem to be coming to a breaking point where wide spread networks and full accessibility seem to be the key words for the present – and more excitingly – for what will happen in the future.

Cities today and the construction of new models

Cities evolved into complex systems, and although it would be possible to analyze different urban models created throughout history and by different civilizations, this would involve writing a whole encyclopedia. The aim of this essay is study the facts about how cities have been created and how tools and productivity allowed them to become the biggest and most complex creation manufactured by mankind.

The modern model of cities, mainly during the last century, has been considerably shaped by the power of the economy. The ability to build networks, infrastructures, knowledge and inventions, has concentrated around power; this power has been constructed around cultural, technological, economical, political groups playing different roles in the social structure and therefore controlling society. Cities are more than buildings, roads or infrastructures; they are constructed "for" people/citizens. The role of the citizen in the traditional models is reduced to the simple use of the city's spaces in order to move around, work, rest and entertain, which meant creating specific areas for specific urban activities in order to organize and manage them. This obviously influenced the definition of the physical space, as well as the social constitution of urban settlements. Today the traditional city model is going through a crisis, as is the economy, politics and many other dimensions of man's everyday life. The lack of formality in the urban growth phenomena has opened new dimensions for studies on how cities could be planned and managed, and even if they should be so, or if they should be considered as living, self-organizing organisms, continuously evolving through time.

The ability to perform tasks and activities without being in a specific location is collapsing the rigid model of cities. Networks of information allow most activities which a single person usually performs in a day to be carried out anywhere. The time table based on: waking up in the morning, taking the train/bus/car to the work/study place, working/studying in a office, stopping at midday to have lunch, working in the afternoon, dinning in the evening, resting and sleeping at home at night, is becoming obsolete. Today all those activities could be performed in a single space with a bed, a table, a chair and a computer with an Internet connection. It is possible to get access to information and products and to perform various tasks remotely using the Web, without having to use money for this purpose (at least not bills or coins). Libraries, offices, classrooms, among thousands of other urban spaces and equipment will need to be reconfigured and readapted for the future needs of citizens. This will also be the case with transport systems, which will need to be rescaled and reshaped for a more efficient performance both to meet the new rules of time adaptability and also to minimize environmental risks from a fossil fuel economy.

The role of the citizen is no longer attached to a professional career or to a single job description lasting his/her entire life. In order to develop intelligent cities, citizens will have to become active and conscious agents in its production, supported by infrastructures capable of generating synergies between them and their own reality. Supported by both software and hardware platforms and therefore networked with peers with common interests and goals, citizens will change their status of user or consumer of the city to become *prosumers*[1] of it.

Today and tomorrow

As a society we might be experiencing one of the biggest social shifts in the last centuries, maybe even comparable with the Enlightenment which represent the start of the last 300 years of history in the Western world. This has had a global effect, changing the position of the human in relation with his environment, society and (him/ her) self. The changes we are experiencing today are not coming from books; they are mainly coming from the vast exchange of information through media, and more specifically through the Internet.

It's an ethic that defines what the new Web is becoming: a massive playground of information bits that are shared and remixed openly into a fluid and participatory tapestry. Having matured beyond its years as a static presentation medium, the Web is now the foundation for new dynamic forms of community and creative expression. Throw in a healthy dose of grassroots entrepreneurship and you have a potent recipe for economic revolution – a revolution that affects not just the obvious targets such as media, entertainment, and software, but is increasingly sweeping across all industries and sectors as mass collaboration makes inroads into activities ranging from science to manufacturing [Tapscott and Williams 2006: 36].

The Internet was born as a military strategy tool to maintain communications between key points in case of a nuclear attack by the URSS on the interests of the United States during the Cold War. The mutation of the Web turned it into the most major agent of change in the last centuries, perhaps comparable only to the steam machine in the Industrial Revolution. The Internet has been the mother ship of the latest revolution that lead us from the "Industrial Society" to the "Information Society", and which is now evolving into the "Knowledge Society". To name these models seems to be a political tool to help define a strategy for what societies are moving towards. The role of governments and politicians shows that a centralized model trying to understand (and drive) the rest of the world, will become and remain obsolete in a world where a distributed and non-hierarchical network of peers will be exchanging knowledge and producing tools and goods faster than ever before.

We are moving towards a different production model, which is at the same time part of a different logical conception. Today, knowledge is not owned by single individuals, nor is it concentrated in major centers filled with gurus and post-PhD people, it relies on collaboration and networks and is evolving and spreading every second. This is only possible using the Internet as a platform (even though the vast majority of the population on earth is not connected to the Web). This has allowed tools (both software and hardware) to grow and mutate at much higher speeds than they used to, and at the same time is connecting them coherently to reality, through the development of user-friendly and accessible applications and products. Wordpress is a blogging platform (software), which allows anyone anywhere to become a website builder and news publisher, in a couple of hours. Arduino is an electronic platform (hardware), used to prototype interactive electronic projects that bring physical computing skills to non-technical people. These two tools are just simple examples among the thousands that can be found today. All the source code for these tools and the projects developed with them is available for everyone to download, modify or adapt it to his/her needs and then upload again. Only a few mouse clicks separate download of the application needed, linking it to your reality and using it as you want. In the book *Wikinomics*, Tapscot and Williams refer to the evolution of tools and technical skills through a very simple example:

He [Jon Petersen] likens developments in the new Web to those of the early history of cinema. "There was a period of time where cinema was a very technical art. You practically had to be an engineer just to run a camera." As the art form evolved, directors stepped up to become storytellers who were less preoccupied with cinematic engineering and more concerned with crafting rich and engaging experiences. "I think something like that is happening on the Web today", says Petersen [Tapscot and Williams 2006: 43].

Accessibility to tools is changing the landscape of knowledge generation and sharing. This is an increasing source of major inventions that are affecting people's everyday life. These sources are not located in universities, companies or research centers; they are "located" in a network of servers linked by the Web. They are generating knowledge and solutions through different platforms where you can find (and support) products and projects that later you will be able to download or develop by yourself. In the business world platforms like Kickstarter[2] are becoming more useful than government grants or private funding for people who want to do applied research or develop products. It allows peers to use the power of the network to fund projects based on crowd-sourcing collaboration. Kickstarter already changed the way people do business, changing the perspective of the entrepreneur: instead of going to the nearest bank in town to ask for a 1,000,000 Euro loan for project development, a more reliable resource is getting 1 Euro from a group of 1,000,000 online people. Tapscot and Williams refer to the power of the many:

As people individually and collectively program the Web, they're increasingly in command. They not only have an abundance of choices, they can increasingly rely on themselves. This is the new consumer power. It's not just the ability to swap suppliers at the click of a mouse, or the prerogative to customize their purchased goods (that was last century). It's the power to become their *own* supplier – in effect to become an economy unto themselves. No matter where one looks these days there is a powerful new economy of sharing and mass collaboration emerging where peers produce their own goods and services. If anyone embodies this new collaborative culture, it's the first generation of youngsters to be socialized in an age of digital technologies [2006: 46].

Digital natives[3] do not have to adapt: the changes happening and about to happen, the concepts of Internet, sharing, open source, networks is a reality for them, as for previous generations the radio, or the TV, among others were.

From Fab Labs to Fab Cities

In digital fabrication the ability to work with digital materials, which can be programmed according to parameters of design, function, ornament, efficiency, etc., is understood. What is now called "digital fabrication" is not quite there yet: digital information is used to operate computer-controlled machines, which can run codes specifying coordinates and instructions like subtract, add or deform materials in different physical states. The road to total digital fabrication passes through the possibility of programming matter at scales which are not visible to the human eye – for example, within the human body and in nature within living organisms – so as tap into the search for energy efficiency, survival or evolution, to mention only some of the characteristics which motivate them.

A path towards digital fabrication has been started, it is taking place within Fab Labs alongside other small-scale workshops worldwide. Now, if we make an analogy with the

origin of the universe, a "Big Bang" period is being lived. We will not however have to wait billions of years to reach our goals. Maybe it will take less time than it took for information to be shared digitally. Taking into account that our great "explosion" happened about ten years ago with the birth of the first Fab Labs in India, Boston and Norway, and where we are today, it is realistic to expect that it will only be a matter of years and not decades until these goals are met. Making an analogy with computers, Neil Gershenfeld refers to how a few decades ago people were working:

> Mainframe computers were expensive machines with limited markets, used by skilled operators working in specialized rooms to perform repetitive industrial operations. We can laugh in retrospect at the small size of the early sales forecasts for computers; when the packaging of computation made it accessible to ordinary people in the form of personal computers, the result was an unprecedented outpouring of new ways to work and play. ... like the earlier transition from mainframes to PCs, the capabilities of machine tools will become accessible to ordinary people in the form of personal fabricators (PFs) [2005: 3].

Fab Labs[4] are fully equipped laboratories (or workshops) with computer-controlled machines that allow anyone to make (almost) anything: from integrated circuits boards to complete houses (i.e., Fab Lab House; see fig. 1), led by Neil Gershenfeld (Center for Bits and Atoms at MIT) and a network of enthusiasts around the world (more than sixty laboratories in over twenty countries today). The Fab Labs are a living project; they are based on a shared knowledge network which distributes processes and projects on the Internet. The difference between this and similar projects already in existence is that the network is based on new physical production methods and their evolution into accessible tools to bring manufacture to the level of the individual. On the Fab Central website we read:

> Fab labs have spread from inner-city Boston to rural India, from South Africa to the North of Norway. Activities in Fab labs range from technological empowerment to peer-to-peer project-based technical training to local problem-solving to small-scale high-tech business incubation to grass-roots research. Projects being developed and produced in Fab labs include solar and wind-powered turbines, thin-client computers and wireless data networks, analytical instrumentation for agriculture and healthcare, custom housing, and rapid-prototyping of rapid-prototyping machines.

Fab Labs today are equipped with CNC machines such as 3D printers, laser cutters, milling machines, knife cutting machines and sufficient equipment to produce in-circuit boards, which allow the production of electronic projects. This vast set of tools allows the production of new technologies, and of course it is possible to produce the machines of a Fab Lab inside a Fab Lab: it can have "children", a self-reproducing organism. The third stage of the Fab Lab road map is the development of micron-scale digital assemblers, which will make it possible to play with materials and operate with their properties to build nanostructures or complex systems with fully integrated intelligence. Even further beyond this, these materials would be capable of being disassembled and reassembled again and again, like a Lego© system. Finally, the fourth stage is in the field of programmable matter, being able to construct things without the need for any kind of machine, emulating nature and the ability it has to reproduce itself, evolve and find functionality without external agents, and with no waste effects.

The roadmap of the Fab Labs:

- 1.0 machines in a small-scale workshop;
- 2.0 A Fab Lab - Fab Lab makes another Fab Lab;
- 3.0 Micron scale material assemblers;
- 4.0 Programmable matter without machine - Digital Fabrication.[5]

Today we are witnessing the birth of the Fab Lab 2.0, which can be reproduced by itself, and its being connected with the worldwide movement of Do-It-Yourself machines, embodied in examples like Rep Rap, MakerBot, Fab @ Home, MTMSnap, among others. It must be understood that, although originating from different fronts, this DIY movement is opening the same road maps in different spaces, garages or research centers: personal fabrication, using the technologies have developed over the last decades by the masses, taking into account the evolution of personal computers, tablets, mobile phones and the Internet. Technical capabilities are being unveiled by content available online and access to production tools, which can be also "downloaded" and assembled in a living room.

Nowadays the Fab Lab network shares content, knowledge and processes online through Internet platforms and live videoconferences. It has allowed the development of projects like the Fab Lab House[6] in Barcelona (fig. 1), or the FabFi[7] in Afghanistan, as well as the constitution of educational platforms for advanced professional training such as the Fab Academy, a shared educational program run by the Fab Lab Network. In the Fab Academy the planet is the campus and the classrooms are the Fab Labs of the world, and content is broadcast by professors from different educational and research institutions (fig. 2).

Fig. 1. Fab Lab House: Project developed by IAAC for the Solar Decathlon 2010 competition, in collaboration with the Center for Bits and Atoms at MIT and the Fab Lab Network. The house was fully fabricated in the Fab Lab Barcelona, and its shape responds to the maximization of radiation surface in Madrid, based on the principle that form follows energy. A team from twenty-five different countries participated in the construction of the prototype, which was the winner of the People's Choice Award at the competition

Fig. 2. Barcelona Interactive model. The Interactive model is a project of IAAC and Fab Lab Barcelona, in collaboration with Hangar (electronics advisor). The interactive model was developed during the IAAC Global School, an intensive workshop with twenty-two participants from five continents in Barcelona; and eight participants from India, at the Balwant Shelth School of Architecture, and in collaboration with Hangar in Barcelona, and the Politecnico di Torino, Italy

Fig. 3. Urban Feeds: Personal data collection devices. A project developed at the Fab Lab Barcelona to make available to citizens the creation of their own environmental sensing devices, being able to collect data, share it and distributed in social networks

---

Besides the technological development of the Fab Labs and other production centers, there is a clear effect that will change the way cities are inhabited and how we interact with them. People in today's society have been given incomparable power of action through mass distribution of knowledge and facilitated access to tools. Not only the Web but also physical space as well is being shaped by the different sets of tools available today. Software is changing the way we move and socialize in the urban space: chat rooms are more populated than many coffee shop in our cities; GPS-enabled hardware and software are modifying our transportation paths; mobile apps are giving us real time information about parking spaces or bikes available in bike sharing services within our cities; the urban landscape is being shaped by non-physical obstacles and by human behavior where it is affected by this new set of tools (fig. 3).

Again, the tool is shaping the city, and vice versa. Providing this new set of tools with the possibility of being changed and produced and not just used, will definitely change and redefine cities. The relationship between the digital and physical worlds becoming intertwined will bring unprecedented consequences for humanity. However, the process of fabrication and raw material sourcing will need to be redefined; environmentally speaking it is not possible for everything to be made by just anyone using traditional material resources. It won't be sustainable to produce furniture in Santiago de Chile with wood sourced from Finland, or to make circuit boards in Chicago with electronic components coming from China. The challenge is in finding new raw materials, not only originating from limited natural resources but also coming from reused materials or new materials being created out of waste or non-used objects. We need to shift from the Planned Obsolescence[8] model, where products need to last for a certain period of time in order to keep the industries producing, which in turn ensures a salary for the workers and the continuation of the economy. We have long been dependent of this model, and it is now bringing us to one of the biggest crises since the great depression of 1929.

FabCity[9] is a new model for the city, which relies on the power of giving back to the cities the ability to produce through micro factories inserted in the urban fabric and connected to the citizens (fig. 4). FabCity relies on the model "from PITO to DIDO".[10] PITO stands for "Product In, Trash Out", the conventional model of a city that has been produced until now: a city that consumes goods, and produces waste and does not gain anything from it is unsustainable at different levels (economic, environmental, social, and/or cultural). DIDO stands for "Data In, Data Out", a city model in which there is no real waste, waste is a resource in itself, making possible a loop of sustainable production and reuse. This is how the city becomes an organism that will bring commodities to people and establish the platforms for knowledge management and sharing, an attractor of talent and exporter of solutions, maximizing its resources.

In practical terms the FabCity will give rise to a new model for the city which redefines the use of new information technologies and production, giving a social, economic, and productive dimension to the tool. The same tool that has been used to construct spectacular sculptures will be reoriented now to offer solutions to local problems: energy, production or socialization of objects. FabCity is a model that is based on the Fab Labs and other platforms and works as a productive and talent-attracting center, located inside the districts of Barcelona. It will be capable of dealing with the realities of ordinary people, but at the same time will be connected to a metropolitan and global network of knowledge related with the use of technologies of "digital fabrication".

Fig. 4. Digital Slums. IAAC and the Universidad Central de Venezuela (UCV). Based on the principle of self-fabrication of slums, the prototype is a reflection on the relation between the low techonology in informal areas and the high technology in the Fab Labs. The prototype was developed in three weeks with forty students from UCV, and is intended to be constructed in Petare (Caracas), one of the largest slums in Latin America

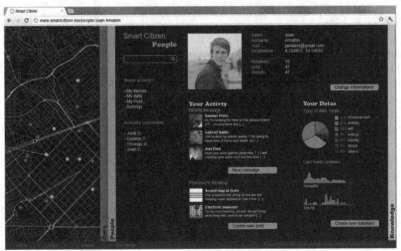

Fig. 5. The Smart Citizen (http://www.smartcitizen.me) project is based on geolocation, Internet and free hardware and software for data collection and sharing, and (in a second phase) the production of objects; it connects people with their environment and their city to create more effective and optimized relationships between resources, technology, communities, services and events in the urban environment. Currently it is being deployed as initial phase in Barcelona city by Fab Lab Barcelona (http://www.fablabbcn.org), IAAC (http://www.iaac.net) and Hangar (http://www. hangar.org), supported by a crowdfunding campaign in Goteo (http://www.goteo.org)

The future is now, the construction of reality is permanent, we have the tools connected in this global network of creation. The medium must be the means; the ends are perhaps more difficult to interpret, to find and define.

From microcontrollers to cities... produced by citizens.

Notes

1. From Wikipedia: "**Prosumer** is a portmanteau formed by contracting either the word **professional** (or less often, **pro**ducer) with the word con**sumer**".
2. Kickstarter is the world's largest funding platform for creative projects. Every week, tens of thousands of amazing people pledge millions of dollars to projects from the worlds of music, film, art, technology, design, food, publishing and other creative fields; see http://www.kickstarter.com.
3. From Wikipedia: A **digital native** is a person who was born during or after the general introduction of digital technology, and through interacting with digital technology from an early age, has a greater understanding of its concepts.
4. From Wikipedia: A **fab lab** (*fabrication laboratory*) is a small-scale workshop offering (personal) digital fabrication.
5. From Neil Gershenfeld's presentation on Fab Labs; see http://ng.cba.mit.edu/show/10.12.fab.show.html.
6. The Fab Lab House was a project developed by the Institute for Advanced Architecture and the Fab Lab Barcelona for the Solar Decathlon Europe 2010 competition. The house was totally fabricated in the Fab Lab Barcelona, and developed by a international team of architects, designers and fabricators worldwide.
7. **FabFi** is an open-source, city-scale, wireless mesh networking system. It is an inexpensive framework for sharing wireless Internet from a central provider across a town or city. It was developed originally by FabLab Jalalabad to provide high-speed Internet to parts of Jalalabad, Afghanistan, and designed for high performance across multiple hops.
8. From Wikipedia: **Planned obsolescence** or **built-in obsolescence** in industrial design is a policy of deliberately planning or designing a product with a limited useful life, so it will become obsolete or nonfunctional after a certain period of time.
9. Antoni Vives introduced the FabCity project in the Fab7 conference held in Lima, Peru, in August 2011. FabCity consists of a network on production centers in the inner city of Barcelona, networked among themselves and serving as knowledge and entrepreneurship platform for citizens.
10. A concept proposed by Antoni Vives (Deputy Mayor of Barcelona City Council), Neil Gershenfeld and Vicente Guallart and supported by worldwide network of Fab Labs.

References

BAUMAN, Zygmunt. 2006. *Vida líquida*. Barcelona: Paidos.
CASTELLS, Manuel. 1996. *The Rise of the Network Society*. Malden, MA: Blackwell Publishers.
GERSHENFELD, Neil A. 2005. *Fab: The Coming Revolution on Your Desktop – From Personal Computers to Personal Fabrication*. New York: Basic Books.
KOOLHAAS, Rem, Stefano BOERI, Sanford KWINTER, Nadia TAZI, and Hans Ulrich OBRIST. 2001. *Mutations*. Bordeaux: ACTAR.
MITCHELL, William J. 2003. *Me++: The Cyborg Self and the Networked City*. Cambridge, MA: MIT Press.
PAPANEK, Victor J. 1971. *Design for the Real World: Human Ecology and Social Change*. New York: Pantheon Books.
TAPSCOTT, Don and Anthony D. WILLIAMS. 2006. *Wikinomics: How Mass Collaboration Changes Everything*. New York: Penguin Group.

About the author

Tomas Diez is a Venezuela-born urbanist specializing in digital fabrication and its implications for models of future cities. He currently leads the project Fab Lab at the Institute for Advanced

Architecture of Catalonia (IAAC) and coordinates the Fab Academy program offered by the worldwide network of Fab Labs. He has participated in technological and social projects such as rehabilitation plans for marginal areas of Caracas, the digital manufacturing installation *Hyper-habitat* for the XI Venice Architecture Biennale, the digital fabrication of the first solar house in Barcelona: the *Fab Lab House* for the Solar Decathlon Europe 2010 in Madrid, the launch of Fab Labs in cities including Lima (Peru), Addis Ababa (Ethiopia) and Ahmedabad (India), among others. He is part of the international team that drives the programs Fab Labs and Informalism and Smart Cities by Smart Citizen, both research-oriented towards the use of ITC for the development of citizen-based platforms for the city production. His research focuses on the use of digital tools for the transformation of physical reality to find a more fluid relation between machines and humans. His work relates the conscious and unconscious actions of human beings with the production of immediate reality through the development of platforms and tools for their participation in this process physically and intellectually.

Gabriela Celani

School of Civil Engineering,
Architecture and Urban Design
University of Campinas
Av. Albert Einstein, 951
Campinas, SP, BRAZIL
CEP 13083852
celani@fec.unicamp.br

Keywords: digital fabrication
laboratory; practical instruction;
laboratory pedagogy;
architectural education

Research

Digital Fabrication Laboratories: Pedagogy and Impacts on Architectural Education

Abstract. This paper discusses the role of the new digital fabrication laboratories in architectural education, as an opportunity to introduce practical exploration along with scientific content. It includes a historical review of practical instruction in architecture, a description of digital fabrication labs, and a comparison between pedagogical methods in engineering laboratories and in digital fabrication labs. The paper ends with a reflection about the impact of the introduction of this type of labs on architectural education.

Introduction

Since the late 1990s a new type of laboratories started to appear in the leading architecture schools, initially called "rapid prototyping labs", and later referred to as "digital fabrication labs".[1] But, aside from the use of digital technologies, how do these laboratories differ from the traditional model shops, and why are they called laboratories, instead of simply "digital fabrication workshops"? Should they focus mainly on research, development or instruction? Which are the pedagogical methods used to teach in these labs and how they can be integrated into the architectural curriculum? Finally, how are these laboratories related to other practical instruction spaces, such as the studio and the science labs, and what can be the consequences of their introduction in architecture schools?

To answer these questions, this paper presents a review of practical instruction in architectural education in Western history, and looks at the types of laboratories and the educational methods for practical instruction used in the neighboring field of engineering. The aim of this work is to help architecture schools define the objectives of their digital fabrication labs, and to propose a way of incorporating practical instruction in these labs within the curriculum, with adequate pedagogical methods in each phase of the learning process. The paper ends by discussing the possible changes that may be introduced in the architectural practice as a result of this new way of teaching.

A review of practical instruction in architectural education

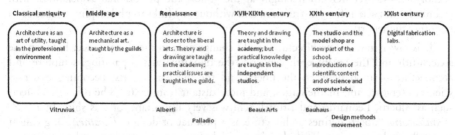

Fig. 1. Practical instruction in architectural education over time

Nexus Netw J 14 (2012) 469–482
DOI 10.1007/s00004-012-0120-x; *published online* 25 September 2012
© 2012 Kim Williams Books, Turin

Fig. 2. Three examples of practical instruction in different periods: in a medieval guild, in Jean Louis Pascal's independent architecture studio in the nineteenth century, and in a Bauhaus workshop in the early twentieth century

Fig. 3. Digital fabrication lab: students gather around the laser cutter; removing a model from the 3D printer; assembling a model with rapid-prototyped parts

In classical antiquity the arts were divided into different categories. The liberal arts, which were exclusively intellectual, were only practiced by free citizens (thus the name "liberal"), while all the other arts, which involved manual work, were practiced by slaves [Jaeger 1986]. Aristotle separated the "arts of necessity" from the "arts of pleasure". Architecture and the figurative arts were not considered part of the superior arts; they were considered arts of necessity, and thus were taught in professional practice, outside of the great philosophers' academies. One of the first initiatives to change this situation was Vitruvius's treaty, *De architectura libri decem*, which would influence the formation of the "modern concept of a broad liberal arts education" of architects [Vitruvius 1999].

In the first universities, in the Middle Ages, the liberal arts taught in the Trivium and the Quadrivium were, respectively, grammar, logic and rhetoric, and arithmetic, geometry, music and astronomy. The "arts of necessity" were then called "mechanical arts", and included blacksmithing, navigation, agriculture and hunting, as well as medicine and architecture (the exact list of mechanical arts varies according to time and author). These arts were then taught at the guilds, the professional associations, with technical – not scientific – instruction, and without any links to philosophy, science or to the other higher arts.

This lower status of architecture would change only in the Renaissance. Between the sixteenthth and the seventeenth centuries the figurative arts – painting, sculpture and architecture – stood between the liberal and the mechanical arts, becoming closer to science, literature and mathematics, and more distant from crafts. The role of architects such as Alberti, Leonardo and Brunelleschi was very important for this shift. In *De Re Aedificatoria*, Alberti defines architecture as a product of design (*lineamentis*), giving it an intellectual dimension. With drawings – including the recently invented perspective – and scale models, architects gained control over the production of the building, and with

treatises like Alberti's and Palladio's they were freed out from the guilds' masters, and architecture became a generalizable science, with its own grammar and theory.

However, the academy was not created to substitute the apprenticeship system, but rather to introduce two novelties: a theoretical discussion about art and architecture, and the drawing as a way to define the building *a priori*. Practical instruction continued to be provided by professional associations and private workshops for many centuries. At the École de Beaux Arts in Paris, for example, studios would be incorporated in the school only in the second half of the nineteenth century.

In the twentieth century, after World War I, the Bauhaus in Weimar was probably the most successful example of the incorporation of the professional workshop in the academic education of architects. Its pedagogy was rooted in the traditional apprenticeship system, and proposed re-integrating technical and aesthetic issues through intense work in production shops: "crafts work was seen as an ideal unity of artistic design and material production" [Bauhaus Dessau Foundation 2011]. Each workshop had two masters: a master of form, "an artist responsible for the design and aesthetic aspect of work", and a master of crafts, "a craftsman who passed on technical skills and abilities".

Between 1923 and 1928, László Moholy-Nagy established what is nowadays considered the essence of the Bauhaus method, promoting the integration of art, science and technology. He was an enthusiast of the qualities of materials and tried to derive aesthetic value from the new industrial production techniques. His ideas were later published in *Von material zu architektur* (1929) (translated into English as *The New Vision* [1938]), which would become an important reference for the modern design method. In his book he proposed the "merging of theory and practice in design" [Moholy-Nagy 1938: 5] and described his educational method:

Teachers and students in close collaboration are bound to find new ways of handling materials … materials through actual experience of its properties, its possibilities in plastic handling, in tectonic creation, in work with tools and machines such as is never attained through book knowledge in the usual school exercises and the traditional courses of instruction [1938: 23].

Although the Bauhaus was a very influential example in architectural education, after World War II many professional schools were incorporated into larger universities, and had to introduce more scientific content in order to gain scientific respectability, as described by Herbert Simon:

It is ironic that in this century the natural sciences almost drove the sciences of the artificial from professional school curricula, a development that peaked about two or three decades after Second World War [1998: 111].

In engineering education, for example, science and mathematics were gradually added to the curriculum. Soon "the practical aspects of engineering generally taught in the laboratory began to give way to the more academic, theoretical subjects" [Feisel and Rosa 2005: 122].

Like engineering schools, architectural programs were also gradually transformed through the inclusion of scientific content, losing part of their traditional hands-on educational methods. This process culminated with the Design Methods Movement in the 1960s, in which the architectural design process was extensively studied in a scholarly way. New "scientific methods" of design were developed and tested, in response to the new, highly complex architectural programs. The movement was influenced by

developments in the fields of operational research, artificial intelligence and computer technology. Its main objectives were 1) to design better, by understanding the process of design; 2) to externalize the design process, allowing large teams to collaborate; 3) to allow repetitive parts of the design process to be automated by the computer [Gregory 1971].

A new type of practical instruction was thus added to architectural education – the science laboratory – which can be defined as:

> a place where scientific research and development is conducted and analyses performed … [using] a vast number of instruments and procedures to study, systematize, or quantify the objects of their attention. Procedures often include sampling, pretreatment and treatment, measurement, calculation, and presentation of results … (*Encyclopaedia Britannica*, 2011).

In the traditional architectural studio and model shop, work was usually not carried out in a scientific, systematic way. Design was usually exploratory, based on trial and error. The new science laboratories in architecture schools were used for the demonstration of physical concepts in architectural applications, such as acoustics, lighting and statics, but also for the scientific study of the design process. It is not clear when and in which field the denomination "design laboratory" first appeared, but nowadays many architecture departments include a laboratory with this name, where research about the design process is conducted.

Another type of laboratory that was introduced in architecture schools in the past decades was the Computer-Aided Design (CAD) lab. In the 1970s the first CAD-specific courses were offered in computer labs; in the 1980s computers started being installed in the studios for 3D modeling, and in the 1990s they were already very common in the studios, with an emphasis in advanced visualization [Mitchell 1990]. At that point, the popularization of 3D modeling and rendering made physical scale models almost disappear in many schools.

In the same way that computers became part of the design studio in the 1980s and 1990s, in the 2000s computers became part of the model shop. Since the late 1990s some schools started introducing rapid prototyping and other computer-controlled machines in what would be later called "digital fabrication laboratories". Initially, these facilities included just smaller rapid prototyping machines for producing scale models automatically from CAD models, but soon there was an interest in the new post-industrial methods for the production of full-scale prototypes and building parts.

Digital fabrication laboratories in architectural education

The introduction of digital fabrication technologies in architectural education is a recent, yet quickly-implemented phenomenon. As stated by Rivka Oxman and Robert Oxman "fabrication labs in education, which were rare even just a few years ago, are today commonplace" [2010: 23]. However, the origins of these laboratories in architecture schools have not been well documented.[2] The first experiences in the use of digital fabrication for architectural prototyping and model-making were the result of the collaboration with mechanical engineering laboratories, which were already using these techniques for product development. As early as 1994, Mitchell and McCullough were already proposing the use of "rapid prototyping" and "numerically controlled fabrication" for making scale models and for producing building parts, directly and indirectly:

…increasingly, it is feasible to use rapid prototyping devices to generate physical scale models from digital information [1994: 461].

Rapid-prototyping machinery can be used not only for direct transformation of CAD models into fabricated objects, but also to produce moulds and dies needed to reproduce those objects in other materials or in multiple copies [1994: 432].

But acquiring computer-controlled machines is not enough for turning a model shop into a laboratory. The essence of a laboratory is in the scientific approach to experimental work, which necessarily includes systematization, the use of control variables, the elaboration of conjectures, and the documentation of all processes.

One of the first digital fabrication laboratories in an architecture school was set up by Professor William Mitchell at MIT's School of Architecture and Planning in the late 1990s. The laboratory's first acquisition was a fusion deposition modeling (FDM) machine, followed by a laser cutter. The first applications of these machines were producing scale models related to Ph.D. students' researches and to graduate elective subjects specially created for exploring the new techniques. Water jet cutters and computer numerical control (CNC) router machines were also available through an agreement with the Mechanical Engineering Department. As the methods and parameters for using the machines were gradually assimilated, a larger number of students started using them, more machines were acquired, and the laboratory became the Digital Design Fabrication Group (http://ddf.mit.edu), where nowadays many courses are taught and research projects are conducted.

The methods, parameters and examples of applications developed by pioneer digital fabrication labs like MIT's made it feasible for other schools to implement their own labs. As the word spread, the price of machines decreased, and new, less expensive techniques became available, architecture schools started creating their own digital fabrication laboratories.

An important aspect of these new resources was their motivational power, as acknowledged by Marc Schnabel in an eCAADe round table on CAAD education, in 2001:

3D Modellers, 3D Scanners, immersive Virtual Environment and Rapid Prototyping are used to assist both students and teachers to explore and study architectural creativity in a new way that enables a deeper involvement into design-issues. Since production time and cost are fairly eliminated, students do not become too attached to a design, which is the outcome of long training of IT-applications, modelling and production. A solution can not only be altered as quickly as new ideas emerge but also experienced virtually or real [Mark, Martens and Oxman 2001: 210].

The "ideal computer curriculum" proposed by Mark, Martens and Oxman included, in the advanced level, a "Computer Aided Manufacturing and Robotics" course, which included "the possibilities of numerical control processing, rapid prototyping and building component manufacturing" [2003: 170]. In the same paper the authors discuss two different strategies for introducing new technological contents in the architectural curriculum: one that integrated digital design topics in existing courses, and another one in which the topics were offered in specific, mandatory courses. The authors concluded that the first strategy was more efficient because "it is essential that an architect's education continue to be focused on issues of building and place".

However, it is still necessary to provide instructions for students regarding the use of the machines and their specific software, and to introduce digital fabrication concepts, such as the three types of computer-controlled production processes – additive, subtractive and formative [Lennings 1997] – because this knowledge can have an impact on design decisions. For this reason a better option may be to include specific courses on digital fabrication in the beginning, and progressively integrate this content in more advanced architectural design studios.

Types of laboratories: research, development and education

Feisel and Rosa [2005] define three types of engineering laboratories, with different objectives: research, development and education. According to them, **research laboratories** "are used to seek broader knowledge that can be generalized and systematized", thus contributing to the overall knowledge in a field. In research projects the use of the digital fabrication laboratories is often combined with the use of equipment from other science labs, such as wind tunnels. An example of this type of research is described by Prata-Shimomura et. al. [2011]. In order to study the characteristics of natural ventilation in different types of urban patterns, Prata-Shimomura produced a scale model of an urban area using a laser cutter. The precision of the model was crucial to the success of the experiment, since a difference of just a few millimeters could change the results in a 1:1000 scale completely. The technique allowed the precise installation of Pitot tubes and anemometers in some surfaces of the building models, through the automated punching of holes, for measuring wind pressure and turbulence. The understanding of urban ventilation gained with this project can be applied in many future urban design projects.

The objective of **development laboratories** is to obtain experimental data to guide professionals in designing and developing products. This type of laboratory is also used to gather specific measurements of performance, "to determine if a design performs as intended" [Feisel and Rosa 2005: 121]. One of the first uses of digital fabrication equipment for product development was carried out by Frank Gehry's architectural firm in the project of a fish sculpture in Barcelona, in 1992. Shelden [2002] describes how the design team used a laser cutter to produce models for testing the fitting and assembly of the parts, and identifies this moment as a key point in the digital revolution in architecture. Digital fabrication labs can also be used for developing building systems and materials. Loveridge [2011], for example, describes the development of a new aluminum foam material in collaboration with materials and fabrication researchers. The material will not be used uniformly throughout objects; the idea is to use different densities, parametrically defined, in areas that are more and less subject to stress.

According to Feisel and Rosa, while the objectives of research and development laboratories are clear, the objectives of **instructional laboratories** "need to be better defined through carefully designed learning objectives" [2005: 121]. It is possible to say that the same applies to digital fabrication labs. The learning objectives in these laboratories are seldom made explicit. The present cost of rapid prototyping and digital fabrication equipment limits the size and number of pieces of equipment that an architecture school can acquire and maintain. For this reason digital fabrication labs often serve simultaneously as the three different types of laboratories described above. Research, development, instruction and even the production of scale models for traditional design courses often overlap in these spaces. Thus, most digital fabrication labs in architecture schools cannot be classified as a specific type of laboratory.

In order to increase the possibilities, many digital fabrication labs from architecture schools have established partnerships with industry and carry out design (development) projects with students (instruction), using scientific methods to derive new knowledge (research). Vogiatzaki-Spiridonidis, for example, has created

a consortium of schools of architecture and small to medium sized enterprises activated in the area of digital manufacturing ... [to] establish an exchange of experiences, information knowledge and expertise regarding the contemporary trends and specific demands of architectural creation between teachers, researchers and enterprises [2009: 5].

The information transmission in this type of partnership works in both ways: students get closer to the production actors in the real world, and at the industry gets in touch with the outcomes of the research being carried out at the university. In projects like this education, research and development activities are completely integrated.

Pedagogical methods in laboratory instruction

An important issue regarding laboratory instruction is related to the **pedagogical methods used**. As early as the first decades of the twentieth century John Dewey, one of the greatest American educators, proposed the introduction of experimental work from children's to adults' education, as a means to revolutionize education. However, in order for his proposition to be effective, he emphasized the need for using scientific methods:

I see at bottom two alternatives between which education must choose if it is not to drift aimlessly. One of them is expressed by the attempt to induce educators to return to the intellectual methods and ideas that arose centuries before scientific method was developed. ... Nevertheless, it is so out of touch with all the conditions of modern life that I believe it is folly to seek salvation in this direction. The other alternative is systematic utilization of scientific method as the pattern and ideal of intelligence exploration and exploitation of the potentialities inherent in experience [Dewey 1997: 85-86].

Dewey also stated that his emphasis on the scientific method did not mean that specialized techniques needed to be used necessarily. He affirmed that scientific method "provides a working pattern" for experiments, but educators should be sensible enough to adapt these methods to different situations, subjects and level of maturity of the students. This is what makes experience effectively educative.

Another important issue in experimental education is related the the objectives of the educational laboratory, which need to be clearly defined. In a colloquium organized in 2002 by the Accreditation Board for Engineering and Technology (ABET), the American accreditation organization for programs in applied science, computing, engineering, and technology, some of the objectives of the engineering instructional laboratories were outlined [Feisel and Peterson 2002]. These included instrumentation to use tools, working with models, learning to do experiments in safety, developing psychomotor capabilities and sensory awareness, collecting and analyzing data, designing and assembling systems, developing creativity and a sense of ethics, and practicing teamwork and communication skills.

In design studios recently taught in digital fabrication labs, students are usually shown working with CNC routers and 3D printers, making models, designing and assembling building systems, working in teams and presenting the results to reviewers, thus meeting most of the objectives that a laboratory should have. Although these studios are unquestionably motivating for students and can produce good results, some of the

objectives of instruction laboratories listed above are usually not made explicit, such as the scientific methods used in the design process.

The different instruction styles in science laboratories are the theme of an interesting and lively debate among educators and psychologists. Dickey and Kosinski [1991] affirm that the traditional laboratory pedagogy is not effective because it is too similar to traditional classroom methods, which are based on the transmission of information. To solve this problem, they propose the concept of "inquiry labs", in which students have to plan and perform their own scientific investigations, instead of blindly following a cookbook. Sweeney and Paradis emphasize the role of "scientific inquiry labs" in the preparation of future researchers:

> Science is, by its nature, a hands-on, inquiry based discipline. Students are unable to fully appreciate the scientific method and the essence of scientific inquiry unless they have the opportunity to acquire and analyse data first-hand [2004: 195].

However, this type of approach may not work in all levels. Kirschner et al. [2006] have shown that guided instruction is also important, based on the differences between the cognitive loads that can be absorbed by expert and novice students. Controlled experiments are limited and do not necessarily stimulate the use of creativity, but they are necessary in courses for beginners. The "minimally guided approach" (also known as "discovery learning, "problem-based learning", "inquiry learning", "experiential learning" and "constructivist learning") is more efficient when used with intermediate and advanced students.

In a study about engineering education, Sheppard et al. [2009] proposed the categorization of laboratory instruction in three levels (fig. 4). For novice students, laboratories are typically used to gather data related to physical evidence in order to contextualize theory. Students in this level must follow the instructor's directions strictly, step by step, in order to reach the desired results, which will demonstrate a concept. In physics courses, for example, classes typically consist of an introductory lecture in which a theoretical principle is introduced and demonstrated. Next, students must do some exercises in order to understand the mathematical description of the theory. The last step consists of developing laboratory simulations that illustrate the same phenomenon. In these structured or "controlled experiments" students can validate the concepts learnt by testing them with different parameters and conditions.

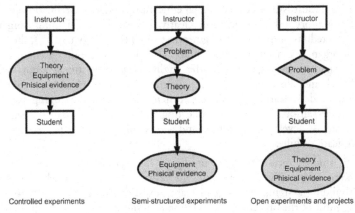

Fig. 4. The three models of laboratory instruction; adapted from [Sheppard et al. 2009]

This initial guided instruction approach is not very commonly used in digital fabrication laboratories (nor in architectural education in general), and could be the reason why students find it difficult to develop their design experiments in a more systematic and scientific way, often recurring to trial and error.

Fig. 5. 3D printer experiment

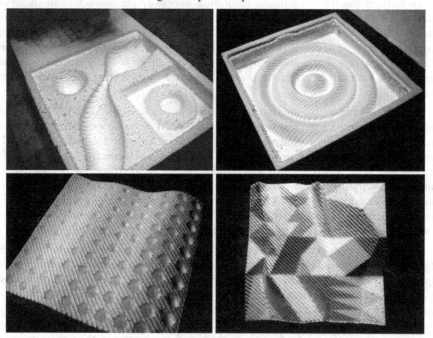

Fig. 6. CNC router experiment. LAPAC, University of Campinas

In digital fabrication labs basic knowledge about the different production methods can be taught in introductory workshops, in which students are asked to develop simple models to explore the specific capacities of each machine. Fig. 5 shows an exercise developed by students from an introductory digital fabrication course taught by Celani and Pupo in which they were asked to develop a 10 x 10 x 5 cm model to be fabricated in a 3D printer, in which they used parametric software to experiment with different wall-widths and feature dimensions. Fig. 6 shows a similar exercise, in which students worked on a standard block of material, varying the type and diameter of tools and experimenting with different milling strategies (such as spiraling, diagonal or orthogonal toolpaths) and parameters (such as vertical and lateral step distances). Different textures could be obtained with the very same geometric model. The result of the milling process with different parameters could be previewed virtually before being sent to the CNC router. With this type of exercises students learn to define parameters and gain confidence in the use of the different machines available in the lab.

The next step in laboratory instruction, still according to Sheppard et al. [2009], consists of allowing students to solve practical problems. In this case students are told which concept to use, but they need to decide on how to use the available equipment to perform a given task. This type of exercise is known as "semi-structured experiment", because only concepts and objectives are given, but no methods are suggested. Semi-structured experiments are more motivating than controlled experiments, because they involve problem-solving, which results in intellectual satisfaction. They make students more pro-active and more confident in the use of concepts, but in order to perform this type of exercise students must have been previously taught how to use laboratory equipment properly.

To work in this level in a digital fabrication laboratory the instructor can challenge students to develop a geometry that can only be produced with a specific type of machine. For example, students can be asked to decide how to build a small ball inside a perforated sphere, or to define the best digital fabrication method for producing a model of a spatial structure. Each student or team will come up with a solution, and some of them will be found to be unpractical, which generates an interesting discussion [Duarte et al. 2011].

A good example of this intermediate approach is described by Mark [2003]. In this design studio students were first taught to write G & M codes for generating CNC tool paths. Next, they had to produce models with three different methods: using CAD commands to generate the geometry and standard CAM software to automatically generate toolpaths, which required a trial-and-error process to achieve the desired results; using CAD scripting to generate the geometry, thus having more control over the toolpaths automatically generated by standard CAM software; and programming directly the CNC pathways as part of the design process, which allowed "greater control over how the forms are shaped and how the materials are fabricated" [Mark 2003: 339].

With advanced students it is possible to use a much more open technique. "Open experiments" can be used to teach how to deal with complex problems. In this type of experiment only concepts and a brief description of the problem are given. Students must define the objectives of their experiment and the means to achieve them. Often they also need to search for new concepts which have not been already introduced by the instructor [Sheppard et al. 2009], or to use interdisciplinary knowledge. Once students are familiar with different production methods and machines and their characteristics, they can develop architectural designs taking these into account. Plus, by this point they

have developed a sense of scientific systematization of procedures, which can help them be more efficient in their design explorations.

An example of this approach can be found in another course taught by Mark [2007]. In this final year architectural design studio students were asked to develop a lightweight oceanfront structure for seasonal use in a coastal island. The structure had to be retractable and covered by a tension membrane, taking into account wind forces and with minimal environmental impact. After defining a preliminary design, students worked with animation software to simulate the transformations of the structures. Some students used CAD scripting to generate their shapes. Next, they used finite element analysis software to simulate the fabric movement and stresses, which helped in selecting the best option for the moveable parts. They then built a physical model in the digital fabrication laboratory to study the movement of the structure and the collision between movable parts. After this, changes were introduced, new computer analyses were carried out, and new physical prototypes were produced. This process was repeated several times, until the project was considered finished. Finally, students used fluid dynamics software for an initial study of the structures' performance, followed by wind tunnel testing with digitally fabricated models. According to Mark, the shift between digital and physical representations, which was facilitated by the use of digital fabrication techniques, was the most important characteristic of the studio:

> As design moves increasingly from paper documentation to computer mediated direct fabrication of architectural projects, greater opportunity exists to associate visual representations on a computer with more dynamic and physical modeling methods. The initial development of a project may involve a wide search of design schemes that seem plausible when simulated with special effects tools. This technology doesn't catch, however, the full range of specific problems of construction, degrees of movement and interference checking realized in rapid prototyping [2007: 226].

In this example, since the problem proposed was very open, students developed significantly different structures, with different types of movement. However, all the projects had in common a scientific process of form optimization through the use of different analysis tools, including digital fabrication, wind tunnel and computer programs.

Discussion: The impact of digital fabrication laboratories on architectural education

We have seen a historical review of practical instruction in architectural education, and the evolution of the field from an exclusively practical profession to the status of a respectable science. Through history, architecture was progressively transformed from a lower art into a highly theoretical discipline. From a hands-on activity, design became a prescriptive activity, in which models and drawings are used to foresee reality, and in which everything must be resolved before the construction process.

Next, we have looked at how practical instruction was swept out and then reintroduced in the professional careers in the twentieth century, and how scientific methods were introduced in architecture, transforming its nature to an even greater extent. We have described the characteristics of digital fabrication labs, and seen how they can be introduced into the architectural curriculum. Finally, we have described the educational objectives and pedagogical methods of science laboratories' instruction, divided in three levels of progressively more open experiments, and we have presented

examples that illustrate how digital fabrication labs can effectively implement the concept of a scientific laboratory and be at the same time a place for creative exploration.

In some schools, the initial impact of digital fabrication laboratories in architectural education was simply the increase in the number of physical models produced [Duarte et al. 2011]. The reason for this was probably the fact that digital fabrication was introduced in specific courses, and not as part of the design curriculum. It is important that students learn to use the machines and understand the concept behind each different digital fabrication strategy in specific courses; but in senior years digital fabrication content should be incorporated in architectural design studios and combined with other technologies, such as parametric modeling, CAD scripting, programming and the use of equipment from other science laboratories, such as natural light simulators, wind tunnels, and computational analysis tools.

As digital fabrication labs become more common in architecture schools and are assimilated by design instructors, they can promote changes in architectural education, allowing students to become closer to the production process and to have a better control over building parts and materials.

Oxman and Oxman [2010] have suggested that the recent interest in fabrication techniques is related to a "cultural shift" in the order in which buildings are defined in contemporary architecture. According to them, in the modern tradition the design process started with the definition of form by the architect alone, followed by the definition of the structure and the material in collaboration with engineers. In a recent phenomenon they call "the new structuralism" material and structure have acquired greater importance in the design process, with form emerging as a consequence of working with the right material in the correct way. In this method structural engineers are present from the very beginning of the design process, working side by side with architects. They describe the recent emergence of interdisciplinary research groups, such as the Arup Advanced Geometry Unit and Smart Geometry, who have developed new design methods with a scientific approach, based on advanced mathematical and computational techniques.

In this scenario, the authors see a new challenge to architectural education: "How do we educate architects to function as material practitioners?" [Oxman and Oxman 2010: 23]. They point out the need to redefining the knowledge base of the architect, which now must necessarily include advanced geometry and "digital enabling skills", and they acknowledge the role of digital fabrication laboratories in this new educational agenda: "Fabrication is not [just] a modeling technique, but a revolution in the making of architecture" [2010: 23].

In summary, it is possible to say that digital fabrication laboratories have a potential of promoting experimental methods in architecture together with a scientific approach, which is the basis of contemporary architecture practice. To achieve this, the explicit use of scientific methods should be encouraged. In fact, this is the reason why they are called "digital fabrication **laboratories**" and not simply "digital fabrication **workshops**". We just have to wait now to see what kind of architects will come out from these new schools, and what type of changes they will impose on architecture.

Acknowledgments

I would like to thank FAPESP, CAPES, CNPq and SAE for funding research, development and instruction at LAPAC, the Laboratory for Automation and Prototyping in Architecture and Construction at the University of Campinas. I would also like to thank all my students and advisees for their diligent work at LAPAC. I would also like to thank Regiane Pupo for her attentive proofreading of the manuscript and Leandro Medrano for his thoughtful suggestions.

Notes

1. The reason for this change is probably related to the fact that mechanical engineers define rapid prototyping as a layered manufacturing process [Volpato 2007], thus restricting the use of this expression to the use of additive machines, such as 3D printers. Other authors, such as Lennings [1997], define RP as "a process that automatically creates a physical prototype from a 3D CAD-Model, in a short period of time" [1997: 297].

2. A search for the keywords "laboratory", "rapid prototyping" and "digital fabrication" in the *Journal of Architectural Education* databases, for example, did not return any entry.

References

BAUHAUS DESSAU FOUNDATION. 2011. www.bauhaus-dessau.de. Last accessed 17.07.2012.

DEWEY, J. 1997. *Experience and education* (1938). New York: Touchstone.

DICKEY, J. L. and R. J. KOSINSKI. 1991. A Practical Plan for Implementing Investigative Laboratories. Available in: http://www.ableweb.org/volumes/vol-12/10-dicke/10-dicke.htm#Intro. Last accessed 17.07.2012.

DUARTE, J. P., G. CELANI amd R. PUPO. 2011. Inserting computational technologies in architectural curricula. Pp. 390-411 in *Computational Design Methods and Technologies: Applications in CAD, CAM and CAE Education*. Ning Gu and Xyangyu Wang, eds. IGI Global.

FEISEL, L. and G. D. PETERSON. 2002. A Colloquy on Learning Objectives for Engineering Educational Laboratories. 2002 ASEE Annual Conference and Exposition, Montreal, Ontario, Canada. Available at: http://mosfet.isu.edu/classes/EE275%20Intro%20Digital%20Lab/Engineering%20Labs%20ASEE%2002%20paper.pdf. Last accessed 17.07.2012.

FEISEL, L. and A. ROSA. 2005. The role of the laboratory in undergraduate engineering education. *Journal of Engineering Education* **94**, 1: 121-130.

GREGORY, S. 1971. State of the art. *Design Methods Group Newsletter* **5**, 6/7: 3.

JAEGER, W. 1986. *Paidéia: A formação do homem grego*. São Paulo: Martins Fontes.

KIRSCHNER, P. A., J. Sweller and R. E. Clark. 2006. Why Minimal Guidance During Instruction Does Not Work: An Analysis of the Failure of Constructivist, Discovery, Problem-Based, Experiential, and Inquiry-Based Teaching. *Educational Psychologist* **41**, 2: 75-86.

LENNINGS, A. F. 1997. CNC offers RP on the Desktop. Pp. 297-301 in *Prototyping Technology International 1997 annual report*.

LOVERIDGE, R. 2011. Parametric Materiality. Pp. 165-174 in *Proceedings of CAADRIA2011*, Newcastle, Australia.

MARK, E. 2003. Programming Architectural Geometry and CNC: Advancing A Design Paradigm with Mathematical Abstraction. Pp. 337-342 in *Proceedings of eCAADe 21*, Graz, Austria.

———. 2007. Simulating Dynamic Forces in Design with Special Effects. Pp. 219-226 in *Proceedings of eCAADe 25*, Frankfurt, Germany.

MARK E., B. Martens and R. E. Oxman. 2001. The Ideal Computer Curriculum. Pp. 168-175 in *Proceedings of eCAADe 19*, Helsinki, Finland.

———. 2003. Round Table Sessions on 'Theoretical and Experimental Issues in the Preliminary Stages of Learning/Teaching CAAD'. Pp. 205-211 in *Proceedings of eCAADe 20*, Warsaw, Poland.

MITCHELL, W. J. and M. MCCULLOUGH. 1994. *Digital Design Media*. New York: John Wiley and Sons.

MOHOLY-NAGY, L. 1938. *The New Vision: Fundamentals of Design, Painting, Sculpture, Architecture*. New York: Norton & Company.

OXMAN R. and R. OXMAN, eds. 2010. *The New Structuralism: Design, Engineering and Architectural Technologies AD Architectural Design* **80**, 4. London: Wiley.

PRATA-SHIMOMURA, A. R., G. CELANI and A. B. FROTA. 2010. Construção de modelos físicos para análise da ventilação em túnel de vento. *Fórum Patrimônio – Clima Urbano e Planejamento das Cidades* **4**, 2. Available at: http://www.forumpatrimonio.com.br/view_full.php?articleID=175&modo=1#. Last accessed 17.07.2012.

SHELDEN, D. R. 2002. Digital surface representation and the constructibility of Gehry's architecture. Ph.D. thesis. Massachusetts Institute of Technology.

SHEPPARD, S. D., K. MACATANGAY, A. COLBY, W. M. SULLIVAN and L. S. SHULMAN. 2008. *Educating Engineers: Designing for the Future of the Field*. Stanford: Jossey-Bass.

SIMON, H. 1996. *The sciences of the artificial* (1969). Cambridge, MA: MIT Press.

SWEENEY, A. E. and J. A. PARADIS. 2004. Developing a Laboratory Model for the Professional Preparation of Future Science Teachers: A Situated Cognition Perspective. *Research in Science Education* **34**: 195-219.

VITRUVIUS. 1999. *Ten Books on Architecture*. I. D. Rowland and T. N. Howe, eds. Cambridge: Cambridge University Press.

VOGIATZAKI-SPIRIDONIDIS, M. 2009. F2F – Continuum Architectural Design and Manufacturing: From the School Lab to the Fabrication Workshop. http://eacea.ec.europa.eu/llp/project_reports/documents/erasmus/erasmus_2007_progress_reports/ecue/eras_ecue_1345 40_continuum.pdf. Last accessed 17.07.2012.

VOLPATO, N. 2007. *Prototipagem Rápida: Tecnologias e Aplicações*. São Paulo: Edgard Blucher.

About the author

Gabriela Celani received her B.A. in Architecture and Urban Planning (1989) and M.Sc. in Architectural Design (1997) from the University of São Paulo, and her Ph.D. in Design and Computation (2002) from MIT, where she was advised by professors Terry Knight and William Mitchell. She is presently a professor of architectural design at the University of Campinas, where she is also the head of LAPAC, the Laboratory for Automation and Prototyping in Architecture and Construction, which she founded in 2007. She is the author of *CAD Criativo* (Rio de Janeiro, Campus-Elsevier, 2003), an introduction to VBA programming for implementing generative design tools, and has translated Mitchell's *The Logic of Architecture* and Moore, Mitchell and Turnbull's *The Poetics of Gardens* into Portuguese. Gabriela is also co-founder and co-editor of *PARC*, an online journal of research in architecture (http://www.fec.unicamp.br/~parc).

Tobias Bonwetsch

ETH Zurich
HIL F56
8093 Zurich, Switzerland
bonwetsch@rob-
technologies.com

Keywords: digital fabrication,
robots in architecture,
CAD/CAM

Research

Robotic Assembly Processes as a Driver in Architectural Design

Abstract. Over the last couple of years industrial robots have increasingly gained the interest of architects and designers. Robotics in architecture and construction has mainly been looked at from an engineering perspective during the latter half of the twentieth century, with the main purpose of automating the building process. Today the focus has turned towards realizing non-standardized designs and developing custom fabrication processes. However, the specific characteristics of the robot, which distinguish it from common computer numerically controlled machines, are often overlooked. Industrial robots are universal fabrication machines that lend themselves especially well to assembly tasks. Applied to architecture this resolves to the ability to control and manipulate the building process. As such, applying industrial robots emphasizes construction as an integral part of architectural design. Moreover, designing and manipulating robotic assembly processes can become a driver in architectural design. The potential of such an approach is discussed on the basis of several design experiments that illustrate that by applying such methods, form is not derived from computation or geometry, but from a physical process.

Introduction

Industrial robots are the latest addition to a number of computer numerically controlled (CNC) fabrication tools, such as routers, mills, or laser-cutters, adopted in the architectural realm over the last years.[1] CNC combined with digital design tools allows the architect to directly transfer design information to fabrication machines. This close coupling of the process of design and making has revived a notion of craft charged with the computational power of the digital tools, which act as stimulation for architectural innovation.[2]

However, applying robotics to construction is not a new concept and has been subject to research for over thirty years, experiencing a notably boom in the 1990s. Spearheaded by Japanese companies and universities, up until now, over 200 different prototypes of robotic solutions have been developed especially for the construction industry and tested on building sites.[3] However, none of them could establish themselves in the industry or passed the prototypical stage.[4] One reason for this might be that instead of exploiting the flexibility of the machine the intention of the research was solely that of automation. Driven by economic factors, the goal was to make construction faster and cheaper, while neglecting the potential for an added value in design quality. The results were, on the one hand, highly specialized machines that automated existing construction processes. On the other hand, attempts were made to fully automate the construction of complete buildings onsite.[5] Considering the specific characteristics of the building industry, where every building is unique, built in accordance to the individual clients needs and for a specific site, both concepts lacked the ability to adapt to varying building challenges and limited the possible design space.[6] Overall the engineering

Nexus Netw J 14 (2012) 483–494

Nexus Netw J – Vol.14, No. 3, 2012 **483**

DOI 10.1007/s00004-012-0119-3; *published online* 27 September 2012

research missed the link to the architectural design and the potential of robots was never made available to the architect.[7] Today, supported by changing technical and economic conditions and a different conceptual approach, the question of how to exploit the flexible potential of robots and strategies on how to integrate this potential into the design process, takes on a greater role.

Robots as programmable design tools

In contrast to earlier endeavours industrial robots are now subject to design explorations.[8] This rediscovery in the field of architecture benefits from changing surrounding circumstances. Today architectural practice is no longer imaginable without the aid of information technology. CAD/CAM already allow for conceiving and realizing differentiated designs in an automated process. Although, articulated arm robots with their six axes are considerably more complex to control than common CNC machines, control systems for robots have become far more sophisticated and thereby easier to program.[9] At the same time, prices of industrial robots are dropping, as manufacturers integrate off-the-shelf personal computer technology and the overall worldwide distribution increases.

But, besides the mere availability, what makes industrial robots of particular interest as a design tool for architects and designers is their universal nature.[10] Robots exhibit specific features which distinguish them from common CNC machines, but which are often overlooked. That is, the programmability of the robot refers both to the digital control of its movement and actions, and to the definition of the physical fabrication process.[11] The latter is not predefined, but dependent on the tool the robot is equipped with. These so called end-effectors can consist of multiple tools or even be changed within a running process (fig. 1). Therefore, industrial robots allow defining the actual material manipulation, which in consequence is subject to design decisions. As the ability to program allows the architect to create his own design tools, partially freeing him from any concepts rooted in a specific CAD software, designing a robotic fabrication process opens up the freedom to follow physical processes outside the given frameworks of common CNC machines.

Fig. 1. A compilation of diverse fabrication processes realized with the same robot, but different uniquely designed end-effectors

Whereas the manufacturing process of common CNC machines is component based – components that still need to be joined and assembled – the kinematics of articulated arm robots lends itself especially well to assembly tasks.[12] This puts robotic fabrication close to actual building practice, as construction can be described as the assembly of different parts and materials. Applying robots as a design tool for architecture results in the ability to control and manipulate the building process (fig. 2).

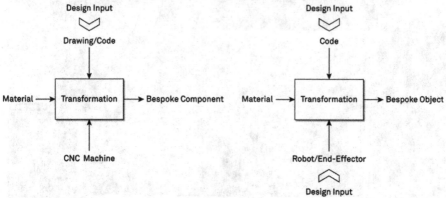

Fig. 2. Diagram of CNC process (left) and robotic process (right)

Programming construction processes

By engineering project specific, or design specific, fabrication processes and thereby giving free rein to the flexibility of the robot, the direct programming of the machine becomes a necessity. In contrast, with common CNC machines applying CAM software to generate machine code from CAD drawings is the norm. However, there are numerous examples of architectural projects where the machine code is directly scripted or generated within the workflow of a digital design process.[13] Nevertheless, the operational step of applying CAM software was skipped in all these projects due to reason of automating, rather than the inability of the CAM software. Especially, when dealing with a large amount of bespoke elements, programming the machine code is more efficient than generating it 'manually' in a CAM software by pointing and clicking. CNC machines are usually self-contained and tightly constrained in the fabrication process they perform, as well as in their working area and position of material. This makes it possible to represent and control their functionality within a general CAM software. At the easiest, fabricating a digitally described object becomes as simple as printing a piece of paper.[14]

Regarding flexible robotic processes, the fabrication process itself is subject to change. The robot can be equipped with completely different tools and peripheral devices and the layout of the working cell can change for each process. This means that, unlike CNC machines, the physical constraints are constantly changing and it would be impossible to foresee all potential processes beforehand in order to represent them in CAM-like software.[15] Therefore, the robotic control code needs to be programed for each specific process and account for its respective characteristics.

The control code for the robot is the encoded description of the process of making. When applying robots for additive fabrication or assembly tasks, the programming of the fabrication process is equivalent to programming the construction process. Assembly basically implies building up a three-dimensional element out of a number of single

pieces that are smaller than the final object. Naturally, the single pieces constituting the final object have to be placed and processed in a certain order to guarantee the feasibility of production. Foremost, the laws of gravity apply. Every piece placed must be supported either by already processed material, or through some kind of scaffolding (fig. 3).

Fig. 3. a, above) Fabrication of wall element. To realize the overhangs a support structure is necessary during the fabrication process; b, below) The completed installation "Structural Oscillation" at the Venice Biennale 2008

Secondly, the reachability of the position of placement must be assured, meaning that if more than one piece could be processed at a given point of time, the one which does not block the other pieces should be processed first. As a consequence, not only does the final structure have to be sound and stable as a whole, but it must be ensured that a stable equilibrium is achieved in each fabrication step during the build-up process.[16] In principle the same rules of common sense apply as in traditional, manual, construction. In addition, the specific anatomy of the robot, its specific end-effector, and the layout of the peripheral components has to be accounted for. This generally implies that manual processes cannot be adopted one-to-one, unless the robot and tool exhibit the same shape and dexterity as the human arm directing the tool; rather they have to be transferred and made suitable for the machine.

Through incorporating the fabrication parameters, programming the robot is the step by step description of the construction process. Of course, one can make use of standard programming techniques using subroutines, loops, conditional statements, etc., in order to structure the code and introduce variations, but in the end the robot has to perform a multitude of steps and actions in a precisely ordered sequence.

Integrating fabrication parameters into the design process

The parameters dictated by the fabrication process – besides the parameters mentioned above these also encompass the specific characteristics of the material and semi-finished products processed[17] –influence the development of a design. They are at once both constraints and opportunities. Once the fabrication parameters are identified and incorporated into the design tools, in this case through code, they open up a design space for exploration.[18] In such a design space, bounded by fabrication constraints, the design exploration is deeply rooted in constructive principles and the physical reality of building.[19]

However, current examples operating in the broad field of digital design and fabrication very seldom choose design strategies that directly incorporate parameters of fabrication. One reason for this might be that, as described above, CAM software allows controlling common CNC machinery through drawings.[20] Drawings as a means of exploring and representation put geometry at the heart of design development. In contrast, the approach proposed of programming construction develops and generates the design out of the fabrication parameters and the necessary sequential fabrication steps.

Recent examples that integrate fabrication parameters into the design process and even apply robotics are the explorations of Brell-Cokcan, Braumann, et al. [2010] and the 'RoboFold' technology developed by Gregory Epps.[21] 'RoboFold' is a sheet metal forming process, where the final form is achieved solely through folding. Possible designs are explored in a design tool that lets the user define curved fold creases and simulates the folding process. Brell-Cokcan, Braumann, et al. apply a robot for 5-axis CNC milling. In particular, flank milling strategies are pursued, thus the process creates ruled surfaces. A parametric design tool allows the user to define and control curves that represent the toolpaths of the milling tool, while incorporating certain relevant fabrication constraints such as tool length and tool diameter.

Although the examples adopt robots for fabrication they do not apply assembly or construction processes, but are limited to formative (i.e., 'RoboFold') or subtractive processes (i.e., milling). As a consequence they can only be applied to fabricate components and to the forming of a single piece.

Applying robotic assembly processes

A number of research projects conducted at the chair of Gramazio & Kohler for Architecture and Digital Fabrication specifically investigate robotic assembly processes. The projects combine both the design and engineering of a robotic fabrication process and consequently the application of the fabrication process on a design task. Design tools developed range from simple scripts to software applications that integrate parameters of the fabrication process (i.e., material and fabrication constraints) for design exploration. In order to broaden the results, several design explorations were carried out within the scope of experimental student courses.

As a basic method, the experiments are built up by first analyzing a manual assembly task and then transferring it to a robotic process. In doing so, aspects in the robotic process that differ from the manual work, as well as the design relevant points of intervention in the digital process control are identified.[22] In a third step the manipulation of these design relevant points, as well as fabrication constraints, are made available in software tools.

The brick projects might act as a simple example.[23] The basic technique of brickwork can be described as piling-up bricks joined by an amorphous connection (e.g., mortar) to create a greater, purposefully shaped whole. In the robotic bricklaying process a gripper attached at the end of the robot-arm places the individual bricks and bonds them with an adhesive. On the one hand, the position and rotation in space of the individual bricks is open for manipulation, while on the other hand, the necessity to follow an ordered construction sequence and limitations in possible placements of the bricks due to the specific shape and design of the gripper act as constraints. The digital control of the process allows the architect to access these parameters and activate them in his design.

For the façade of the Gantenbein winery the bricks were positioned in a rigorous grid and solely the degree of rotation around the center axis of each brick was exploited as a design parameter, thereby allowing the display of a pattern on the façade and at the same time controlling the amount of light entering the building (fig. 4). However, the control for the oscillating wall segments of the installation at the eleventh Venice Architectural Biennale is more elaborate. The complex shape of the elements and with it the position of the single bricks is determined by the constructive requirement that the segments need to be self-supporting and firm. The resulting overhangs can only be realized by introducing support bricks that are integrated in the fabrication process (fig. 3).

More elaborated assembly processes allow optimizing material and structural efficiency through specific localized allocation of material. Furthermore, through the combination of different materials in a single fabrication process hybrid constructions can be realized that implement diverse functional and aesthetic properties. Such concepts were followed in the applied student research project on robotically fabricating outdoor furniture for the ETH campus. The assembly process consists of the layering of foam blocks joined with an adhesive. The blocks feature variable length and differing density and hardness. By means of controlled distribution of blocks with different material properties throughout the cross section it was possible to address functional requirements such as seating comfort, weight and stability. The flexibility of a backrest, for instance, could be influenced by combining blocks of different elasticity. The resulting student designs work as a communicative seating area, a bench or a private armchair, depending on their orientation (fig. 5).

Fig. 4. a, above) A pattern of spheres emerges on the façade of the Gantenbein winery;
b, below) View of the interior space illustrating the light atmosphere accomplished through the
rotation of the individual bricks

Fig. 5. a, above) Foam blocks of different density and hardness are distributed over the cross section in order to control localized performance properties; b, below) The finished furniture pieces are coated with a layer of Polyurethane

Conclusion

The examples exhibit certain characteristics, and show advantages and potentials of programming the construction process and activating the robot as a design tool. The projects have in common that the final geometry is created through the description of the

process of making. The knowledge of making – both construction knowledge and knowledge of the tools applied – is codified and acts as driver of the design. Thereby, design and form are not primarily derived from computation or geometry, but from a physical process.

A very straightforward advantage of design through programming the fabrication process is that every design conceived within the design space spanned by the fabrication parameters is buildable and needs no post-processing. When integrating the logic of construction and the fabrication constraints in the design phase, traditional intermediate steps – those which transfer a design into something buildable – are skipped. Construction design and shop drawings that give instructions to the craftsmen and builders for the purpose of execution are eliminated. The design is thus closely connected to the physical reality of building, thereby reducing transfer loss from conception to construction.

Additionally, using code to describe a design and ultimately the process of its making takes advantage of more general aspects of notation. These are the potential to produce multiple instances of a design, as well as the possibility of inserting variation within a copy of the code. Further, through abstraction of the design, the code makes it possible to insert variables, establish relations, and implement iterations and conditionals. Programming code thus creates the frame for design exploration.[24]

The power of abstraction is facilitated by computers. Foremost, the computer has the ability to manage and process a large volume of data, which the robot can transfer into a physical process. Hence, the digital description of an object can be extremely specific and in consequence the code for fabrication can consist of a myriad of different instructions. As every step of the construction process has to be explicitly defined, the physical manifestation of a design can be highly informed and precisely defined down to its smallest constituent element.[25]

This is especially the case for programming assembly processes. With common CNC machines performing subtractive or formative fabrication processes the information in form of complexity or intelligence is introduced on the surface and the shape of the component. The single components still need to be assembled, a job that can become quite challenging for a large quantity of pieces, especially considering the potential of every component to be uniquely shaped. However, applying the concept of programming the fabrication process to assembly processes already embraces the assembly task, and additionally offers the potential to introduce complexity in the placement of the individual units that form the whole. Assembly processes allow the designer to gain control over the macro- and the micro-structure of a design and the definition of its cross section.

Acknowledgements

This paper draws upon the research work of the chair of Gramazio and Kohler for Architecture and Digital Fabrication, ETH Zurich. The author wishes to thank all people of the group involved, especially Ralph Baertschi, Fabio Gramazio, Michael Knauss, Matthias Kohler, Michael Lyrenmann, and Silvan Oesterle, as well as all the students participating in our teaching courses.

Notes

1. Several design schools around the world have installed robotic fabrication laboratories, including ETH Zurich (http://www.dfab.arch.ethz.ch/index.php?lang=e&this_page =infrastructure), Harvard Graduate School of Design (http://www.gsd.harvard.edu/inside/ cadcam/), Carnegie Mellon University (http://www.cmu-dfab.com/), the Southern California

Institute of Architecture (http://www.sciarc.edu/portal/about/resources/robotics_lab.html), University of Stuttgart (http://icd.uni-stuttgart.de/?p=4052), and others.

2. See for instance [Iwamoto 2009] and [Kolarevic 2008], which collect a number of advanced and innovative projects and experiments in the field of digital design and fabrication in architecture.

3. This number includes various mechatronic devices from completely autonomously working machines to teleoperated apparatus. Also, not all perform tasks that are strictly necessary for construction, as for instance painting. For an overview see [Bock 2008] and [Howe 2000].

4. See [Balaguer and Abderrahim 2008].

5. The concept of a fully automated factory portal was primarily intended for high-rises. They set high constraints on the building design and additionally the typically very dense surrounding of construction sites for high-rises proved a barrier to keep up the supply chain and logistics.

6. For a discussion of the specific characteristics of the building industry see for instance [Groák 1992: 126-127]. These specific characteristics are also the reason why concepts of other manufacturing industries cannot be directly transferred to the building industry, as discussed by Richard Harris and Michael Buzzelli [2005].

7. Furthermore, the research and developments was primarily executed by process engineers. This compassed the special know-how of professional builders and architects in the construction of buildings and its direct relation to the architectural design process; see [van Gassell and Mass 2008].

8. For an overview of robotic driven design experiments conducted by the group of Gramazio & Kohler see [Gramazio and Kohler 2008].

9. Instead of positioning the robot by controlling the angles of each rotary joint, inverse kinematics and advanced controls allow moving the robot in various coordinate systems simply by defining a point in space.

10. In the patent description for a 'Programmed Article Transfer' – considered to be the first industrial robot – George Devol describes the object of the invention to be 'universal automation' and draws a direct analogy to computers. Where the latter is a 'universal machine' for office work – and nowadays almost all aspects of our lives – the former equals for fabrication: a general purpose machine. See [Devol 1954].

11. Many current examples indeed apply robots as a universal fabrication machine, but in a way that they only mimic other CNC machines, without activating the intrinsic potential of the robot. Although interesting projects result, but they are not specific to robotic fabrication and, for instance, might just as well have been done with a 5-axis router.

12. CNC machines have their origin in the automation of machine tools and are therefore developed to machine components, mainly through cutting or deformation. Industrial robots on the other hand, with their dexterity resembling that of a human arm, saw their first assignment in the handling of parts; see [Engelberger 2007].

13. For examples again see the collections for projects in [Iwamoto 2009] and [Kolarevic 2008].

14. Several digital fabrication tools are already integrated as simple printer-drivers within CAD software, such as 2D laser cutting, or 3D stereo lithography printing. Here, fabricating physical objects from a digital drawing becomes as easy as pressing the 'print' button.

15. This is only economically feasible for clearly specified and fixed installed robotic processes, as for instance using the robotic arm as a router.

16. Consider for instance stone arches or shell structures. Although some specific geometries allow for stable configurations during erection without the need for support, forces can only be transferred once the final geometry is achieved and can be activated.

17. If we stick to the simple process of stacking discrete pieces, the friction coefficient of the material or semi-finished product used will, for instance, influence the steepness of the angle at which single pieces can still be stacked on top of one another without shearing.

18. The potential of constraints acting as a design driver that can lead to novel design solutions is also emphasized by Axel Kilian. Although his focus is not specifically set on fabrication, independent of the types of constraints, Kilian argues that computational models and the ability to write problem specific proprietary software plays a crucial role in facilitating design explorations. See [Kilian 2006].

19. In architectural history novel forms of construction always had a great impact on design. There are several examples where architects who were also builders developed their designs directly out of assembly processes, like Brunelleschi in the early Renaissance or modern examples like Pier Luigi Nervi and Felix Candela. See [Groák 1992: 192].

20. On how the architects adopted CNC technology only once they were able to control the machines through drawing, which was made possible through the widespread introduction of CAD/CAM, rather than writing numerical code, see [Callicott 2001: 55].

21. [Tachi and Epps 2011]; see also the company's website: http://www.robofold.com/.

22. The goal of the research projects is not automation, but to investigate architectural design potentials. Therefore, the implementation of the robotic process is focused only on those parts were the digital process control becomes relevant to design.

23. For more details on the brick projects see [Bonwetsch, Gramazio and Kohler 2007] and [Bärtschi et al. 2010].

24. On the advantage of notation cf. [McCullough 1998: 94].

25. On how the digital may effect a new materiality see [Picon 2004].

References

BALAGUER, Carlos, and Mohamed ABDERRAHIM. 2008. Trends in Robotics and Automation in Construction. Pp. 1-20 in *Robotics and Automation in Construction*, Carlos Balaguer and Mohamed Abderrahim, eds. InTech. http://www.intechopen.com/books/robotics_and_automation_in_construction. Last accessed 04.08.2012.

BÄRTSCHI, Ralph, Michael KNAUSS, Tobias BONWETSCH, Fabio GRAMAZIO, and Matthias KOHLER. 2010. Wiggled Brick Bond. Paper presented at the Advances in Architectural Geometry, Vienna, 2010.

BOCK, Thomas. 2008. Construction Automation and Robotics. Pp. 21-42 in *Robotics and Automation in Construction*, Carlos Balaguer and Mohamed Abderrahim, eds. In-Teh.

BONWETSCH, Tobias, Fabio GRAMAZIO, and Matthias KOHLER. 2007. Digitally Fabricating Non-Standardised Brick Walls. Paper presented at the ManuBuild, 1st International Conference, Rotterdam, 2007.

BRELL-COKCAN, Sigrid, and Johannes BRAUMANN. 2010. A New Parametric Design Tool for Robot Milling. Pp. 357-363 in *ACADIA 10: LIFE in:formation, On Responsive Information and Variations in Architecture* (Proceedings of the 30th Annual Conference of the Association for Computer Aided Design in Architecture (ACADIA)). New York.

CALLICOTT, Nick. 2001. *Computer-Aided Manufacture in Architecture the Pursuit of Novelty*. Oxford: Architectuaral Press.

DEVOL, George Charles. 1954. Programmed Article Transfer. United States Patent 2,988,237, filed 12.10. 1954, and issued 6.13.1961.

ENGELBERGER, Joseph F. 2007. Historical Perspective and Role in Automation. Pp. 3-10 in *Handbook of Industrial Robotics*, 2nd ed., Y. Nof Shimon, ed. Wiley Online Library, http://onlinelibrary.wiley.com/doi/10.1002/9780470172506.ch1/summary.

GASSEL, Frans van, and Ger MAAS. 2008. Mechanising, Robotising and Automating Constrcution Processes. Pp. 43-52 in *Robotics and Automation in Construction*, Carlos Balaguer and Mohamed Abderrahim, eds. InTech. http://www.intechopen.com/books/howtoreference/robotics_and_automation_in_construction/mechanising__robotising_and_automating_constr uction_processes. Last accessed 04.08.2012.

GRAMAZIO, Fabio, and Matthias KOHLER. 2008. *Digital Materiality in Architecture*. Baden: Lars Müller Publishers.

GROÁK, Steven. 1992. *The Idea of Building Thought and Action in the Design and Production of Buildings*. London: E & FN Spon.

HARRIS, Richard, and Michael BUZZELLI. 2005. House Building in the Machine Age, 1920s–1970s: Realities and Perceptions of Modernisation in North America and Australia. *Business History* 47, 1: 59-85.

HOWE, A. Scott. 2000. Designing for Automated Construction. *Automation in Construction* 9, 3: 259-76.

IWAMOTO, Lisa. 2009. *Digital Fabrications Architectural and Material Techniques*. Architecture Briefs. New York: Princeton Architectural Press.

KILIAN, Axel. 2006. Design Exploration through Bidirectional Modeling of Constraints. Ph.D. thesis, Massachusetts Institute of Technology.

KOLAREVIC, Branko. 2008. *Manufacturing Material Effects Rethinking Design and Making in Architecture*. New York: Routledge.

McCULLOUGH, Malcolm. 1998. *Abstracting Craft the Practiced Digital Hand*. Cambridge, MA: MIT Press.

PICON, Antoine. 2004. Architecture and the Virtual: Towards a New Materiality? *Praxis: Journal of Writing+Building* 6: 114-21.

TACHI, Tomohiro, and Gregory EPPS. 2011. Designing One-Dof Mechanisms for Architecture by Rationalizing Curved Folding. In *International Symposium on Algorithmic Design for Architecture and Urban Design (ALGODE-AIJ)*. Tokyo.

About the author

Tobias Bonwetsch is senior researcher at ETH's Laboratory for Architecture and Digital Fabrication chaired by Gramazio & Kohler. He studied architecture and graduated from the Technical University of Darmstadt. After gaining working experience as an independent architect, he completed his postgraduate studies at the ETH Zurich with a specialisation in computer-aided architectural design. His research is concerned with integrating the logic of digital fabrication into the architectural design process, with a special focus on additive production methods. In 2010, Tobias Bonwetsch co-founded ROB Technologies, a company that develops and implements robotic manufacturing processes at the interface of architectural design and the building industry.

Aineias Oikonomou

25, Iasonos street
Vouliagmeni
166 71 Athens, GREECE
aineias4@yahoo.com

Research

Design and Tracing of Post-Byzantine Churches in the Florina Area, Northwestern Greece

Keywords: architect's cubit, Byzantine foot, design analysis, Florina area, northwestern Greece, post-Byzantine period, single-nave church, three-aisled basilica, structural systems, symbolism, ad quadratum, ad triangulum, diagonals, geometry, geometric systems, golden section, harmonic proportions, irrational numbers, measuring systems, module, proportion, proportional systems, symmetry

Abstract. This paper constitutes an effort to investigate the application of specific design and constructional tracing in post-Byzantine churches situated in northwestern Greece, mainly in the wider area of the town of Florina and the Prespa Lakes, covering a period from the fifteenth to the nineteenth century. Firstly, a review of the relevant previous studies concerning the investigation of proportions and constructional tracing in Byzantine churches is presented. Secondly, a brief analysis of the metric systems of the Byzantine and the Ottoman period is given in order to define the applied modules. Furthermore, the typology of post-Byzantine churches is analysed. The main part of the paper includes the investigation of the constructional grid and the proportions in several plan drawings, in combination with the use of triangular tracing (right triangles). The analysis is also extended to façades and sections, in order to demonstrate the standardisation of the construction. Apart from the above, the application of the golden section in some examples is also examined. Finally, conclusions are drawn concerning the design and application of models and tracing on the construction of post-Byzantine churches.

1 Previous studies of tracing and proportion

The investigation of proportions and constructional tracing in Byzantine churches has been the object of many previous studies, both in Bulgaria and in Greece. In Bulgaria, concerning medieval churches, I. Popov [1955] investigated the proportions using diagonal tracing and squares (quadratura or ad quadratum), while G. Kozukharov [1974] used equilateral and right-angled triangles in order to explore circular and arched tracing. In Greece, N. K. Moutsopoulos [1956, 1963, 2010] applied geometrical tracing in sections of Byzantine inscribed cross-shaped churches and argued that there exist constructional tracings that define the proportions of their parts. Furthermore, K. Oikonomou [2003] analysed sections of Byzantine churches using diagonal and square constructional tracing (quadratura), while A. Prepis [2007] analysed plans of Serbian and Albanian post-Byzantine churches using diagonal circular tracing.

P. A. Rappoport [1995] analysed the building process of Russian Kievan churches from the tenth to the twelfth centuries, reported specific external dimensions (30x20 'belts' – measuring units) for these Byzantine monuments [Rappoport 1995: 163] and stated that:

> it has been supposed that the working system of a Kievan architect was based on geometrical constructions. However, in the absence of drawings, it is more likely that the dimensions of the building were determined arithmetically, in the form of simple proportional relationships, with linear measures as an initial unit' [Rappoport 1995: 167].

Concerning Byzantine architecture between the ninth and fifteenth centuries, R. Ousterhout [1999] analysed the building methods of Byzantine master builders and emphasised on the use of modular systems based on the Byzantine foot and the existence of specific geometric ratios such as 4:3 and 2:1 [Ousterhout 1999: 80]. He characteristically states that "the application of geometry must begin with a system of measurement" [Ousterhout 1999: 75].

Apart from the above, B. Sisa analysed plans and sections of Protestant wooden bell towers in the Carpathian basin using as a starting point the medieval theory of the constructional principles of the circle-square-triangle,[1] which provides evidence that carpenters did not use drawings but applied these principles directly during the construction stage [Sisa 1990: 327]. Sisa's drawing analysis showed that these timber structures are based on a constructional grid with specified proportions [Sisa 1990: 338, 347].

Based on the afore-mentioned bibliographical review, it can be seen that apart from the studies by Ousterhout and Rappoport, who refer to specific module and dimensions in Byzantine feet, most of the other studies do not use or present a given module, but apply tracing of circles and diagonals which subdivide the façades or sections in smaller parts, without mentioning the use of the module.

This paper constitutes an effort to investigate the application of specific design and constructional tracing in post-Byzantine churches situated in northwestern Greece, mainly in the wider area of the town of Florina and the Prespa Lakes, covering a period from the fifteenth to the nineteenth century. The significance of the present study, in relation to previous ones that refer to the tracing of medieval churches, lies mainly on the detection of a specific construction module and certain metric models which were applied in churches of northwestern Greece, in combination with the use of constructional tracing based on the application of the Pythagorean Theorem.

2 Methodology

The starting point of this study is the author's Ph.D. thesis [Oikonomou 2007], which involved the analysis of the typological, morphological, constructional and environmental characteristics of forty traditional houses in Florina. An important part of this work was dedicated to the investigation of the application of the module in houses of the nineteenth century.[2] The research led to a scientific paper presented in Greece [Oikonomou and Dimitsantou-Kremezi 2009]. Afterwards, post-doctoral research conducted in the Laboratory of Architectural Form and Orders, N.T.U.A. School of Architecture, led to an extended paper presented in Turkey [Oikonomou et al. 2009] and a research article published in *Nexus Network Journal* vol. 13, 3 [Oikonomou 2011]. The research concerned the use of the module, metric models and triangular tracing in the traditional architecture of northern Greece.

The current work is a further extension of the above-mentioned approach to selected post-Byzantine churches in the Florina area. This investigation is based on the application of the constructional module (a modular grid) and triangular tracing to original or redrawn plans and façades of selected churches. A total of seven representative monuments from the prefecture of Florina were selected and analysed. These are situated in the wider Florina area (Vevi), the Prespa Lakes area (Laimos, Platy, Kallithea, Agathoto) and the Greek-Albanian borders (Moschohori and Krystallopigi). The aim of this analysis is to investigate the influence of the module and triangles on the standardisation of the design of the plans and façades, as well as on the construction.

Apart from the above, the plans of eighty post-Byzantine churches of Florina prefecture drawn by Moutsopoulos [2003] were analysed using the same methodology, in order to obtain a better understanding of the standardisation of plans, their proportions and constructional dimensions.

First of all, the typology of specific churches and the metric systems in use are analysed. The older monuments are small single-nave churches and their design is based on the Byzantine metric system (*foot, spithamai* and *orgyia*), whereas the most recent examples are three-aisled basilicas, where the design and construction is based on the Ottoman architect's cubit (*mimar arşin*). Consequently, the first part of the paper also presents these two specific metric systems and the module.

Afterwards, specific metric models that are used in the construction of the designated churches are presented and, at the same time, the use of triangular tracing is analysed. The constructional grid and the proportions of the perimeter are investigated in several plan drawings, as explained above.

The extension of the analysis to the churches' façades and sections shows that there is a standardisation concerning the placement of structural elements, such as the roof trusses and the horizontal wooden ties. It also reveals that constructional tracings exist, but only in the plan and its tracing on the ground, whereas the proportions of façades and heights derive from the consequent subdivisions and not from circular tracing. Nevertheless, it is shown that the design and construction of façades is aided by the use of right triangles.

The final part of the paper draws conclusions concerning the design and the tracing of post-Byzantine monuments, as well as proposals for further research, which may include older examples (before the fifteenth century).

3 Investigation of the application of the module and triangular tracing in the post-Byzantine churches of north-western Greece

3.1 The Byzantine and Ottoman metric systems

The Byzantine metric system was based on two different measures: the Byzantine *spithami* (handbreadth equal to 23.4 cm) and the Byzantine *foot* (31.2 cm). The spithami is divided into 12 *daktyloi* (fingers), whereas the foot is equal to 16 *daktyloi*. It should be noted that 4 spithamai are equal to 3 feet (93.6 cm). During survey and construction the master builders also used the *orgyia* (210.8 cm), which was usually a reed or stick divided into nine spithamai or 108 daktyloi [Ousterhout 1999: 60]. For land measurement, the *schoinion* (rope) equal to 10 orgyai, was used. Ousterhout notices that "the use of ropes suggests a similarity between the process of surveying and the process of laying out a building on-site" [Ousterhout 1999: 60].

The Byzantine metric measures were used by Christian master builders in the construction of churches not only during the whole Byzantine period, but also during the fifteenth and sixteenth centuries (first centuries of Ottoman domination) (see also [Bouras 2001: 242], concerning the similar architectural proportions during these periods).

The measuring of plots and the construction of buildings in the Ottoman Empire was based on the mimar arşin or architect's cubit. The arşin is divided into 24 *parmak*, or fingers; the finger is divided into 12 *hatt*, or lines; and the line is divided into 12 *nokta*, or points [Cerasi 1988: 92]. The arşin is also divided into two equal parts called *kadem* [Özdural 1988: 113].

The Ottomans used the architect's cubit from the beginning of the sixteenth century, though its precise measurement varied. A. Özdural [1988: 103] mentions that the imperial measuring unit was equal to 72.1 cm in 1520; 73.4 cm during the last quarter of the sixteenth century; and 76.4 cm during the third quarter of the eighteenth century. The architect's cubit of 75.8 cm was the fourth and final version, assigned as the imperial measuring unit by Selim III, during the years 1794-95; it remained in use until 1934, when it was replaced by the metre [Özdural 1988: 106].

The arşin, similar to the old English yard, was both an instrument and a measuring unit [Cerasi 1988: 92]. As an instrument, it was usually made of ebony, wood, bronze, copper or bone and could be folded into two (*kadem*) or four parts. As mentioned, it was subdivided into parts of 12 or 24 [Özdural 1988: 113].

3.2 The basic metric models

Based on the analysis of the plans, the monuments are divided into five basic categories:

i.   Churches with plan proportions 3:2, where the tracing is based on two consecutive right 3-4-5 triangles (21 churches);
ii. Churches with plan proportions 4:3, where the tracing is based on one right 3-4-5 triangle (13 churches);
iii. Churches with plan proportions 1.618:1, 21:13 or 13:8, where the tracing is based on the golden section ($\sqrt{5}+1:2$) (5 churches);
iv. Churches with plan proportions 8:5, where the tracing is based on right triangles and circles (5 churches);
v.   Churches with plan proportions 2:1, where the tracing is based on triangles and multiple circles (12 churches).

The metric models, which can be found in the post-Byzantine churches of north-western Greece, as well as in the wider Balkan area, are possibly based on rough plans (*skariphos*) [Ousterhout 1999: 62] that master builders widely used.

In the church architecture of the Florina area, the most common metric models that are observed are:

a. Church with a plan of 20x15 Byzantine feet (ratio 4:3);
b. Church with a plan of 24x16 Byzantine feet (ratio 3:2);
c. Church with a plan of 16x10 arşin (ratio 8:5);
d. Church with a plan of 24x12 arşin (ratio 2:1);
e. Church with a plan of 24x16 arşin (ratio 3:2);
f. Church with a plan of 32x20 arşin (ratio 8:5).

All the above numbers refer to the exterior dimensions of the plan. At this point, a standardisation concerning the interior dimensions should be noted. This is closely related to the standard thickness of the walls, which is 3 spithamai or 1 arşin. For instance, a church with exterior dimensions 20x15 Byzantine feet (app. 26x20 spithamai) and walls that are 3 spithamai thick, has interior dimensions 20x14 spithamai.

Finally, it should be noted that the metric system of the Byzantine foot or the architect's cubit is also applied to the churches' façades, where, similar to the plans, a standardisation of sizes is also observed. The total height of small churches (fifteenth-sixteenth centuries) without the roof (pediment) is usually 9 or 12 Byzantine feet. In contrast, during the eighteenth and nineteenth century, large basilicas with a total height of 9 or 12 arşin are constructed.

3.3 The use of triangular tracing

The metric models that are analysed contribute to the standardisation of the construction of the post-Byzantine church. Most of the churches presented have a fully rectangular plan. At the same time, the constructional module is applied accurately, not only in the placement of the basic wooden structural elements and the openings, but also in the dimensions of the spaces. All the above lead to the conclusion that before the construction of the church, certain basic architectural tracing, which assumes basic knowledge of geometry, took place.

For the tracing of the rectangular perimeter of the church and its main walls, the Pythagorean Theorem with the basic right-angled triangle 3-4-5[3] is applied with the use of knotted ropes. Stoichkov has proven that master builders in Bulgaria used the 3-4-5 triangle during the eighteenth and nineteenth centuries [Stoichkov 1977: 126, 145]. It is also mentioned that "the masters of Bratzigovo ... traced their plans directly on the site, which would sometimes result in harmonic relation and even in golden rule proportions" [Cerasi 1988: 93].[4] Furthermore, Cerasi mentions that "the analysis of plans and elevations of buildings by Balkan masters has shown the existence of proportions and tracing inscribed in circles" and also notes that the use of long leather strips in Bulgaria for the measuring and tracing on the site encourages circular and diagonal tracing [Cerasi 1988: 95].

The same methods certainly apply in northwestern Greece during the eighteenth and nineteenth centuries, but also during the whole post-Byzantine period. Ousterhout characteristically states for the construction of churches in Western Europe that "to lay out a building, one simply had to *know the ropes*" [Ousterhout 1999: 60].

The 3-4-5 triangle can be multiplied by various factors to produce triangles of the necessary dimension; thus multiplication by a factor of 2 produces the 6-8-10 triangle; by a factor of 3, 9-12-15; by a factor of 4, 12-16-20; by a factor of 5, 15-20-25; and so forth. Furthermore, the use of circular auxiliary tracing can not be ruled out, especially in the case of churches that have a plan with a ratio of 13:8. These numbers are related to the subdivision of a line with a length of 13 into two parts with lengths of 8 and 5 (extreme and mean ratio) and the Fibonacci series [see Konstantinidis 1961: 190-191].

The following diagrams include observations concerning the various metric models and their way of tracing (fig. 1). In this way, the relationship with the above-mentioned, characteristic right-angled triangles is investigated and conclusions concerning the possible tracing of the churches' walls and their design can be drawn.

At this point it is useful to cite the definition given by Jouven concerning the (harmonic) tracing:

A harmonic trace is a geometric figure whose design coincides with the main lines of a building (or what amounts to the same, its graphic representation in geometrical projection). The choice of this geometric figure, conceived at the same time as the building by the architect, is based on the remarkable properties that are included in it [Jouven 1951: 7].

Jouven also explains:

We mean by main lines those whose knowledge is sufficient to determine the elements of construction (perimeter, axes, lines of change, etc.). We can say in a quick way that they are the main lines of a first rough drawing of the building" [Jouven 1951: 7].

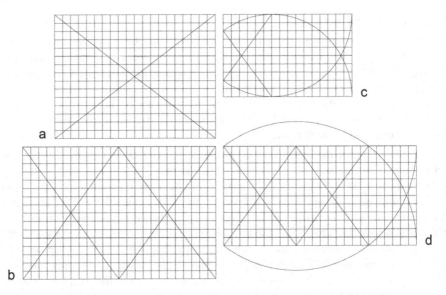

Fig. 1. Use of the module and triangles 6-8-10, 9-12-15, 12-16-20, 15-20-25

We must bear in mind that the tracing of the perimeter and the elements of the church's upper part (dome, roof or upper storey) is also made on the ground, and the construction is based on the specific grid and module (foot or architect's cubit). This can result in a rather complex triangular tracing, without necessarily meaning that the diagonals are in fact traced on the ground: only the corners of the church walls are marked by small stones and then the plan is laid out with knotted ropes [Ousterhout 1999: 60]. A stone is placed the centre of the altar and uncut stones mark the four corners. Then, the master architect traces the plan with the use of a measuring instrument and finally the perimeter of the church is marked with a help of a cord. [Rappoport 1995: 162]

In this way it is impossible to find any sign or mark of the presented tracings as they were made on the ground before the churches were actually built. Nevertheless, master builders often used on the façades circumscribed symbols like the five-pointed star (pentalpha), the six-pointed star (double triangle) and the double square. It is a scientific question if these symbols are in any way related to any possible geometrical tracing on the ground, or if they stand for other purposes.

3.4 Building typology

The basic building type of the post-Byzantine churches of the fifteenth and sixteenth centuries is that of the single-nave church *(monohoros dromikos)*, whereas in the eighteenth and nineteenth centuries, the type of the single-nave church with portico and the type of the three-aisled basilica *(trikliti basiliki)* are commonly used in the wider Balkan area. Christians were not allowed to build churches of great size, nor to use marks and signs such as the cross. As a result most of the post-Byzantine churches have a small size and many of them resemble to houses or farmhouses (see also [Bouras 2001: 267]).

This prohibition of building grand churches by the Christians concerns mainly the towns and villages that were predominantly inhabited by Muslims. Nevertheless, between the sixteenth and the eighteenth centuries, important churches in terms of size and

intentions, of the Athonite cross-in-square type, have been built in monasteries [Bouras 2001: 279].

It was only after the first Ottoman Tanzimat Reform (Hati Serif of Giullhane - 1839) that Christians throughout the Empire were widely allowed to build large basilicas and to use the sign of the cross on the façades, although there are some exceptions built in the eighteenth century, in areas like Voio, Zagori and Pelion [Bouras 2001: 284-285].

4 Presentation of the study

4.1 St. Nikolaos in Vevi (1460)

St. Nikolaos church in Vevi (Banitsa) is a small single-nave church covered with a timber roof, which was built, according to the founder's inscription, in 1460 [Subotić 1980: 86-94].

The analysis of the plan and its comparison with other churches of the same period that can be found in the wider area, reveal a series of important data. The plan is inscribed in a rectangle with proportions 3:2 and mean dimensions around 7.50 x 5.00 m. (24 x 16 Byzantine feet of 31.2 cm). The constructional tracing of the walls is probably based on the use of two right-angled triangles (12-16-20) placed consecutively (fig. 2), while the 6-8-10 triangle (in spithamai) is most likely used for the tracing of the apse and the recess (fig. 3). The mean internal dimensions of the church are 5.40 x 3.50 m (23 x 15 Byzantine spithamai).

It is noted that the tracing includes the three-sided apse, whereas in churches with semi-circular protrusions (see §§ 4.2-4.3), the tracing of the perimeter does not incorporate the recess. The deviation of the northern wall from the traced perimeter is most likely to be attributed to a constructional error. Thus only the southern and eastern walls are perpendicular, whereas the shape of the plan is slightly asymmetrical.

The height of the western façade up to the roof level is equal to 12 Byzantine feet, while the total height including the pediment is 16 Byzantine feet, because the roof height is 4 Byzantine feet. In this way, the western façade is inscribed in a square with dimensions 16 x 16 feet. The design of the façade is aided by the use of two 12-16-20 triangles (fig. 4).

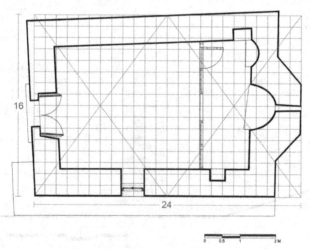

Fig. 2. St. Nikolaos in Vevi. Plan (Byzantine feet grid)

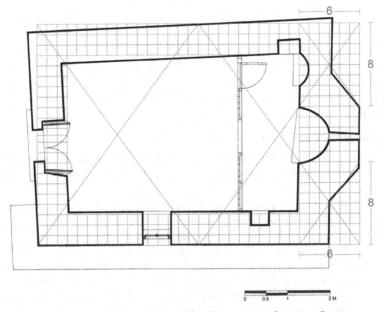

Fig. 3. St. Nikolaos in Vevi. Plan (Byzantine spithamai grid)

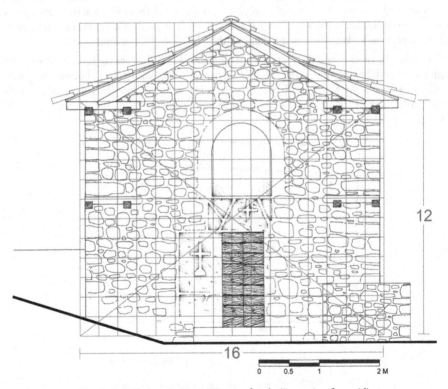

Fig. 4. St. Nikolaos in Vevi. Western façade (Byzantine feet grid)

As far as the construction of the church is concerned, the analysis of its structural elements also leads to interesting findings. The wall thickness is 70 to 75 cm, equal to approximately 3 Byzantine spithamai of 23.4 cm or 1 Ottoman arşin of 72.1 cm (sixteenth century). The wooden ties are placed at a height of approximately 1 Byzantine orgyia of 210.8 cm, while there is a second pair of wooden ties placed at the roof level, at a height of 12 Byzantine feet (375 cm) from the floor (fig. 5). The use of visible wooden ties is typical in small Byzantine and post-Byzantine single-nave churches.

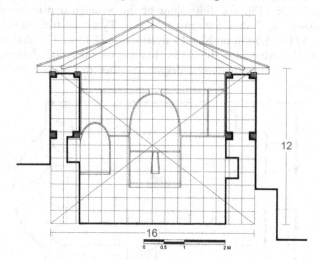

Fig. 5. St. Nikolaos in Vevi. Section (Byzantine feet grid)

The trusses follow a certain module and are placed at approximately 93 cm (3 feet of 4 spithamai) intervals. This standardisation, in combination with the particular internal dimensions of the plan, lead to the conclusion that in St. Nikolaos in Vevi, the original roof must have been composed by nine trusses with a height of 125 cm. This hypothesis is confirmed by the coincidence of these trusses with the zones of the internal paintings on the northern wall. The construction is aided by the use of two circles with radius 12 feet and two triangles 12-16-20 (fig. 6).

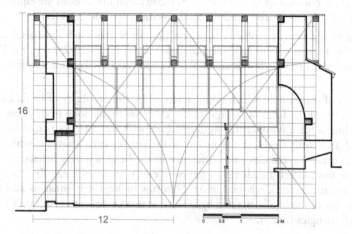

Fig. 6. St. Nikolaos in Vevi. Longitudinal section (Byzantine feet grid)

4.2 Ypapanti in Laimos (fifteenth c.)

The church of Ypapanti in Laimos (Rombi), during its first phase (fifteenth century) [Moutsopoulos 2003: 122], had a plan with similar external dimensions (7.35 x 4.95 m or 24 x 16 feet) to Ag. Nikolaos in Vevi. The tracing of the perimeter is probably based on the use of two consecutive 12-16-20 triangles. The intersection of the diagonals and the triangular tracing coincides with the beginning of the different level of the bema (fig. 7). It is also interesting to note that in another church, that of St. George in Kallithea [Moutsopoulos 2003: 133], the dimensions of the rectangular plan are similar. The external dimensions are 7.45 x 5.00 m or 24 x 16 Byzantine feet, while the internal dimensions are 5.60 x 3.50 m or 24 x 15 Byzantine spithamai. The church of Prophet Elias in Achlada [Moutsopoulos 2003: 188] has also the same plan dimensions (7.50 x 5.00 m or 24 x 16 Byzantine feet), despite the fact that it was built during the seventeenth century.

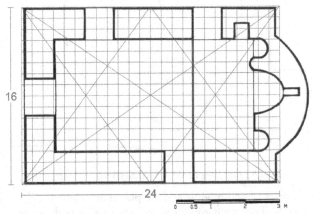

Fig. 7. Ypapanti in Laimos. Plan (Byzantine feet grid)

4.3 St. Nikolaos in Platy (1591)

The church of St. Nikolaos in Platy (Styrkovo) dates back to 1591 [Moutsopoulos 2003: 110-111] and has a plan similar to the above-mentioned churches, except that in this case, the proportions of the plan are 4:3; the external dimensions are equal to 6.20 x 4.70 m (20 x 15 Byzantine feet) and the apse protrudes by 1.00 m (3 feet).

The tracing of the plan is probably based on the use of the 3-4-5 right-angled triangle (multiplied by a factor of 5 to produce a triangle measuring 15-20-25) (fig. 8). The mean internal dimensions of the church are 4.70 x 3.25 m (20 x 14 Byzantine spithamai) (fig. 10).

The form of the church, which is characterised by the simplicity of its facades, is not very different from the form of the rest of the churches that belong to the same building type. The church has a symmetrical western façade with the entrance placed in its centre and a blind arch with traces of the image of the patron saint above it, while the roof forms a pediment. Nevertheless, the difference in the heights is very interesting, because the height of the western façade up to the roof level is in St. Nikolaos in Platy equal to 2.80 m (9 Byzantine feet), whereas in St. Nikolaos in Vevi it is equal to 12 Byzantine feet. The design of the façade is aided by the use of two circles with a radius of 9 feet and two 9-12-16 triangles (fig. 9).

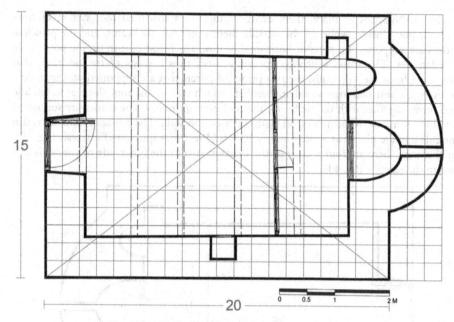

Fig. 8. St. Nikolaos in Platy. Plan (Byzantine feet grid)

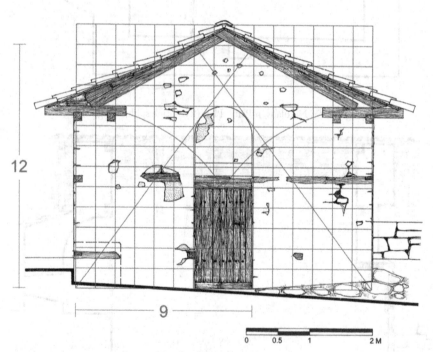

Fig. 9. St. Nikolaos in Platy. Western façade (Byzantine feet grid)

In St. Nikolaos in Platy, there exists an ingenious structural roof system of truss without posts inside, while on the outside there aren't any tie beams, and the principal rafters are supported on parts of horizontal beams, which rest in turn on the protruding wooden ties. The truss, which is repeatedly used, thus consists of the tie beam that protrudes from the side walls, the two principal rafters that are supported on the middle of the walls and two long wooden chamfers, which change the roof slope at its middle and form, together with the tie beam, the roof eaves *(astreha)*. The trusses are supported on the perimetric wooden ties on both sides of the wall. The trusses follow a certain module and are placed at approximately 93 cm (3 feet of 4 spithamai) intervals (fig. 10). In this way, there are four trusses in the interior, two in the inner side of the walls and two in the exterior (fig. 11).

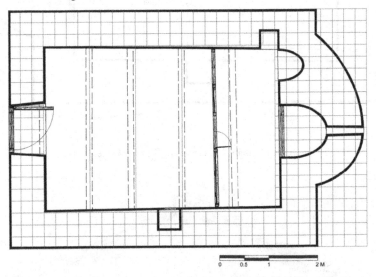

Fig. 10. St. Nikolaos in Platy. Plan (Byzantine spithamai grid)

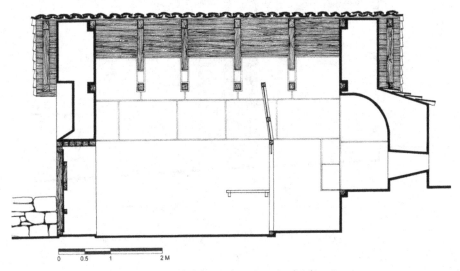

Fig. 11. St. Nikolaos in Platy. Longitudinal section

4.4 St. Athanasios in Kallithea (nineteenth c.)

The church of St. Athanasios is situated on a hill, to the northwest of the village Kallithea (Roudari) in the Prespa Lakes area [Moutsopoulos 2003: 128-129]. It is a small single-nave church with a wooden roof and an inverted L-shaped portico on its western and northern side. It was probably built during the nineteenth century. Two distinct constructional phases can be discerned: first, the nave was built and to this, an outer narthex was added, forming a portico in the southern side of the church.

The external dimensions of the main church are 910x560 cm (12x7.4 arşin). Its constructional tracing is most probably based on the golden section (fig. 12). This hypothesis provides an explanation for the fact that the width of the church is not a rational, but an irrational number $[24/(\sqrt{5}+1)]$. It is also interesting to note the circular rosette with two overlapping five-pointed stars (symbols closely linked to the golden section and the Pythagorean philosophy), on the centre *(omfalos)* of the initial wooden roof of the main church (fig. 13).

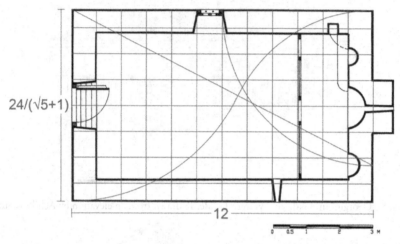

Fig. 12. St. Athanasios in Kallithea. Plan (Arşin grid). Redrawn from a plan by Achilleas Stoios

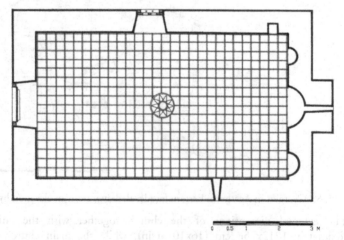

Fig. 13. St. Athanasios in Kallithea. Reflected ceiling plan

4.5 St. Georgios in Agathoto (1869)

The church of St. Georgios is situated in the abandoned village of Agathoto (Tyrnovo), in the Prespa Lakes area, near the Greek-Albanian border [Moutsopoulos 2003: 114-116]. It is a small single-nave church with a wooden roof and an inverted L-shaped portico on its western and northern side, which dates to 1869, according to the stone engraving over the western entrance (fig. 14). On the eastern façade there is a second stone engraving with a cross and a six-pointed geometrical rose (fig. 15). The church was most probably built in one phase.

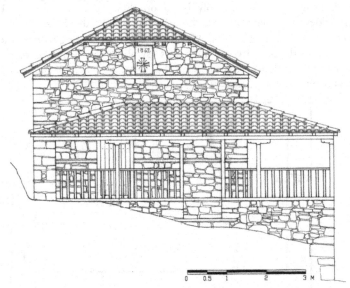

Fig. 14. St. Georgios in Agathoto. Western façade. Redrawn from an elevation by Achilleas Stoios

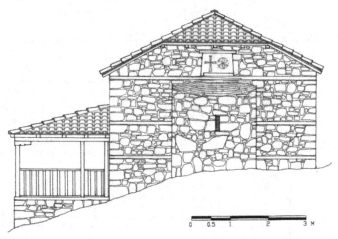

Fig. 15. St. Georgios in Agathoto. Eastern façade. Redrawn from an elevation by Achilleas Stoios

The external dimensions of the church together with the outer narthex are approximately 1212x758 cm (16x10 arşin), while the main church has dimensions 985x530 cm (13x7 arşin). The constructional tracing of the church perimeter including

the portico, is based on two circles with a radius of 10 arşin (vesica pisces) and two 6-8-10 right-angled triangles (fig. 16). The triangles help to trace the western entrance, while the division of the length into 6 and 10 arşin coincides with the interior partition of the place set apart for women. The main space of the church is traced with the use of a circle with a 7-arşin radius.

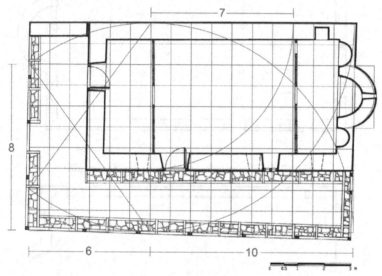

Fig. 16. St. Georgios in Agathoto. Plan (Arşin grid). Redrawn from a plan by Achilleas Stoios

4.6 St. Dimitrios in Moschohori (1871)

The settlement Moschohori (Vambeli) is situated near the Greek-Albanian border, at a distance of fifty kilometres from the town of Florina. St. Dimitrios is a large church, pertaining to the type of the three-aisled basilica, which was built, according to the stone lintel of the entrance, on 1871 [Moutsopoulos 2003: 310]. Its roof has sustained serious damage, which resulted in its collapse on 1965, leading in turn to the destruction of the church's interior. The church had a raised place set apart for women (*gynaikonitis*), located in its western part. The quality of the stonework and the joint-fillings leads to the conclusion that the church was built by an important company of builders (*esnafi*), probably coming from the villages of Voio or those of Epirus. The openings have stone frames and stone arches on their upper part. On the western façade, there existed a stone cross at the upper part, and on the entrance a portico, supported by stone pillars. On the eastern façade, there existed a circular rose window with a twelve-pointed star and two stone cypress trees on its sides, on the pediment over the apse. All these stone parts have fallen in the interior of the church, along with many of the wooden columns that supported the roof and formed the aisles. In the interior of the church, an important part of the mortars and finishes are saved, while the zones of the mezzanine and the raised part for women can also be discerned.

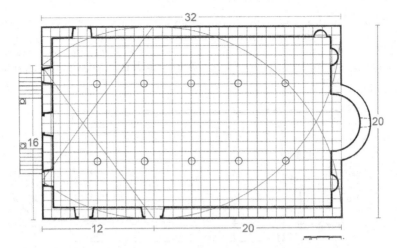

Fig. 17. St. Dimitrios in Moschohori. Plan (Arşin grid)

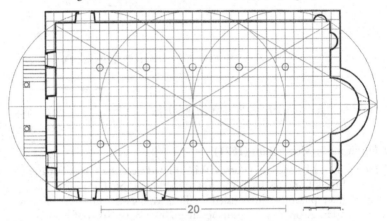

Fig. 18. St. Dimitrios in Moschohori. Plan (Arşin grid)

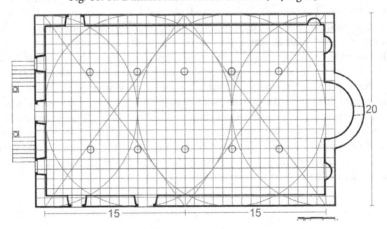

Fig. 19. St. Dimitrios in Moschohori. Plan (Arşin grid)

The plan of the church has external dimensions of approximately 2390x1494 cm (32x20 arşin), while the tracing of the external perimeter is probably based on the use of two circles with a radius of 20 arşin (vesica pisces) and two 12-16-20 right-angled triangles (fig. 17). The centre of one of the circles coincides with the middle of the main, southern entrance, while in the eastern part the circles define the points of the genesis of the apse recess. The tracing of the walls is most probably based on the use of three circles with a radius of 10 arşin (ad triangulum),[5] because their section defines the internal side of the walls (fig. 18). The triangle that is formed in the eastern part also defines the recess. At the same time, two consecutive 15-20-25 triangles and circles with a diameter of 20 arşin define characteristic points of the plan, as the columns and the internal perimeter (fig. 19).

4.7 St. Georgios in Krystallopigi (1891)

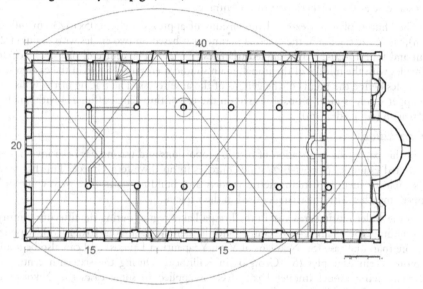

Fig. 20. St. Georgios in Krystallopigi. Plan (Arşin grid)

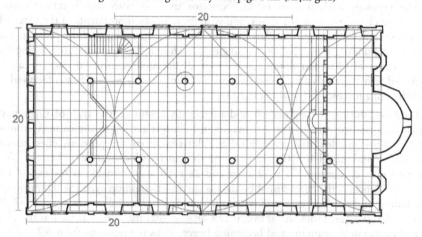

Fig. 21. St. Georgios in Krystallopigi. Plan (Arşin grid)

Krystallopigi (Smardesi) is situated near the Greek-Albanian border, at a distance of approximately fifty kilometres from the town of Florina. St. Georgios church is situated east of the existing settlement, in the place of the initial settlement that was destroyed during the Greek civil war.

It is a large three-aisled basilica with narthex [Moutsopoulos 2003: 309]. In the central axis of the western façade there is a bell-tower. On the same façade, there is a stone engraving with a pediment stating the date, 1891. The church has sustained many interventions on all its facades, which are evident from the different qualities of the stonework and it also has a more recent roof. The openings have stone neoclassical frames, whereas at the corners of the western façade is found a particular form of dressed corner stones, which is reminiscent of Renaissance prototypes. In the interior of the church, there still exists the raised level of the place set apart for women, which is reached by a staircase at the northern part of the entrance.

The church plan has external dimensions of approximately 3005x1523 cm (40x20 arşin), while the tracing of the external perimeter is based on two circles with radius of 25 arşin and four 15-20-25 right-angled triangles (fig. 20). The centre of one of the circles coincides with the middle of the initial, main southern entrance, while in the eastern side the circles define the points of the genesis of the apse recess. The tracing of the walls and the upper storey is most probably based on the use of three circles (fig. 21) with a radius of 10 arşin (ad quadratum).[5]

5 Conclusions

In this paper, the application of metric proportions and tracing has been investigated in selected examples of post-Byzantine churches in northwestern Greece. This has made it possible to identify certain metric models and building types that were commonly applied, together with the triangles used in their tracing.

In the wider Florina area and in the Prespa Lakes region during the fifteenth century, the building type of single-nave church and the same plan (model 16x24 feet) is applied to more than one case (St. Nikolaos of Vevi, Ypapanti in Laimos) and can also be found in more recent examples (St. Georgios in Kallithea). During the sixteenth century, a different metric model (model 15x20 feet) is applied in some cases (St. Nikolaos in Platy), while the dominant building type remains the same.

The typological similarities of St. Nikolaos in Vevi with the fifteenth century churches, such as Ypapanti in Laimos, and with more recent monuments of the sixteenth century, such as St. Nikolaos in Platy, actually reveal the spreading and the prevalence of this certain building type, that of the single-nave church, in the wider area during the first centuries of Ottoman domination. From all the afore-mentioned data, it can be concluded that there existed a certain design model or prototype, which was followed in more than one instance [Mamaloukos 2003].

Later on, during the eighteenth and nineteenth centuries, the type of single-nave church with portico prevails (St. Athanasios in Kallithea and St. Georgios in Agathoto) and the metric models that are applied are different from the previous ones (model 12x7.4 arşin and model 16x10 arşin). During the nineteenth century were built basilicas that have plan dimensions 24 x 12 arşin and 24 x 16 arşin. Finally, at the end of the nineteenth century, the three-aisled basilica is the most common type (St. Dimitrios in Moschohori and St. Georgios in Krystallopigi) and the plans follow the models 32x20 arşin or 40x20 arşin. From the above, it can be seen that the size – as well as the plan ratio – is constantly changing and becoming larger. Thus the ratios go from 3:2 and 4:3

(fifteenth-sixteenth centuries), to 1.618:1 and 8:5 (eighteenth-nineteenth centuries), to 2:1 (end of the nineteenth century).

Another important finding is that the tracing of post-Byzantine churches of the fifteenth and sixteenth centuries, is based on the Byzantine foot and spithami, whereas in the majority of the other presented examples (eighteenth and nineteenth centuries), the Ottoman arşin is used. This rather important finding is explained by the fact that during the first centuries of Ottoman rule, the Christian master builders carried on the Byzantine building tradition, while the Balkan master builders of later centuries were occupied also in other important buildings as mosques and Ottoman baths (hamams).

Concerning the use of triangular tracing, it should be noted that during the whole post-Byzantine period, the basic triangle 3-4-5 and its transformations (multiplied by factors of 2, 3, 4 and 5) is used for the tracing of the perimeter. Nevertheless, changes in the building types and the dimensions of the plan resulted in significant changes in the use of orthogonal triangles. During the fifteenth century, two consecutive 3-4-5 right triangles are applied, whereas in the sixteenth century, the perimeter of the plan is traced with the use of one right triangle. In contrast, during the eighteenth and nineteenth centuries, the design and tracing are based not only on the 3-4-5 right triangles, but also on circles. The ad-triangulum[5] system is probably used in these churches. Last, but not least, in some examples (St. Athanasios in Kallithea), the tracing is based entirely on the golden section ($\sqrt{5}+1: 2$) and the resulting width of the plan equals an irrational number.

From the above, it can be seen that the choice of certain basic models/prototypes and triangles is directly linked to the application of the predominant building types during the different periods (small single-nave church, church with a portico and large three-aisled basilica) and largely affects the form and the proportions of the churches.

Further research should include a larger number of monuments from the wider Balkan area, including Byzantine churches of the twelfth, thirteenth and fourteenth centuries, pertaining to different types. In that way, the principles and the proportions of the Byzantine and Post-Byzantine monuments can be derived and similarities and differences between the various types noted.

Acknowledgments

The author would like to thank Arch. Eng. Achilleas Stoios, for providing some of the original plans of post-Byzantine churches of the wider Prespa area, Prof. Stavros Mamaloukos (Dr. Arch. Eng.) for his valuable scientific advice concerning the design and construction of Byzantine and post-Byzantine churches, and also the 16th Ephorate of Byzantine Antiquities in Kastoria.

Notes

1. See also Jouven [1951: 25], for the principle "circle-square-triangle" by the Bauhutten building guild in Germany, according to Ghyka [1931].
2. This investigation departs from the approach described by Cerasi [1988] and Stoichkov [1977].
3. See also Choisy [1976: vol 1, 49-50] and Konstantinidis [1961: 112] concerning the construction of right angles with the application of the Pythagorean Theorem in the Egyptian and Indian architecture, respectively. The 3-4-5 right triangle is also called rope-stretcher's triangle or Egyptian triangle.
4. Cerasi [1988] cites [Berbenliev and Partasev 1963], as well as his personal communications with P. Berbenliev.
5. See also Jouven [1951: 23, 24], for the council concerning the construction and completion of the Milan Cathedral ad quadratum or ad triangulum (1391).

References

BERBENLIEV, P. and V. H. PARTASEV. 1963. *Bratzigovskite Maistori Stroiteli prez XVIII i XIX vek i tahnoto architekturno tvorcestvo*. Sofia: Tehnika (in Bulgarian).

BOURAS, Ch. 2001. *Byzantine and Post-Byzantine Architecture in Greece*. Athens: Melissa.

CERASI, M. 1988. Late-Ottoman Architects and Master Builders. Pp. 87-102 in *Muqarnas V: An Annual on Islamic Art and Architecture*, O. Grabar, ed. Leiden: E.J. Brill.

CHOISY, A. 1976. *Histoire de l'Architecture*, 2 vols. France: Editions SERG.

GHYKA, M. 1931. Le nombre d'or. Paris: Gallimard.

JOUVEN, G. 1951. *Rythme et Architecture. Les Traces Harmoniques.* Paris: Editions Vincent, Freal et Cie.

KONSTANTINIDIS, D. 1961. *Peri armonikon haraxeon eis tin arhitektonikin kai tas eikastikas tehnas. Historia kai Aesthitiki (On Harmonic Tracing in Architecture and Visual Arts. History and Aesthetics)*. Athens. (in Greek).

KOZUKHAROV, G. 1974. *Svodut v Anticnostta u Srednite Vekove*. Sofia: Anticnostta (in Bulgarian).

MAMALOUKOS, St. 2003. Design Issues in Byzantine Architecture. Pp. 119-130 in *Deltion tis Christianikis Archaeologikis Etairias*. Vol. 24. Athens. (in Greek).

MOUTSOPOULOS, N. K. 2003. *Byzantina kai Metabyzantina Mnimeia tis Makedonias. Ekklisies tou Nomou Florinas (Byzantine and Post-Byzantine Monuments of Macedonia. Churches of Florina Prefecture)*. Thessaloniki (in Greek).

———. 2010. *Naodomia (Construction d'Églises)*. Thessaloniki: University Studio Press (in Greek).

OIKONOMOU, A. 2011. The Use of the Module, Metric Models and Triangular Tracing in the Traditional Architecture of Northern Greece. In *Nexus Network Journal* **13**, 3: 763-792.

———. 2007. Comparative Investigation of the Architectural Structure and the Environmental Performance of 19th Century Traditional Houses in Florina. Ph.D. dissertation. Athens, Greece: National Technical University of Athens, School of Architecture (in Greek).

OIKONOMOU, A. and A. DIMITSANTOU-KREMEZI. 2009. Investigation of the Application of the Module in the Traditional Architecture of Florina. Pp. 309-318 in *Appropriate Interventions for the Safeguarding of Monuments and Historical Buildings. New Design Tendencies. Proceedings of the 3rd National Congress*, M. Dousi, P. Nikiforidis, eds. Thessaloniki: Ianos Publications (in Greek).

OIKONOMOU, A. et al. 2009. Application of the Module and Construction Principles in the Traditional Architecture of Northern Greece. Pp. 543-562 in *LIVENARCH IV (RE/DE) Constructions in Architecture. 4th International Congress Liveable Environments and Architecture*. vol. 2. Ş. Ö. Gür, ed. Trabzon, Turkey: Karadeniz Technical University, Faculty of Architecture, Department of Architecture.

OIKONOMOU, K. 2003. *I Architectoniki tis Koimiseos Theotokou stin Aiani Kozanis. Morfologiki-Typologiki Diereynisi (Architectural Remarks on the Church of the Koimises of Virgin Mary in Aiani)*. Thessaloniki (in Greek).

OUSTERHOUT, R. 1999. *Master Builders of Byzantium*. Princeton: Princeton University Press.

ÖZDURAL, A. 1988. Sinan's Arşin: A Survey of Ottoman Architectural Metrology. Pp. 101-115 in *Muqarnas XV: An Annual on the Visual Culture of the Islamic World*, G. Necipoglu, ed. Leiden: E.J. Brill.

PREPIS, A. 2007. Andrej Damjanov and his Contribution in Church Architecture of Western-Central Balkans in the 19th Century. In *Istoria Domikon Kataskeuon (History of Constructions). National Conference*, N. Barkas, ed. Xanthi: Department of Architecture, Democritus University of Thrace (in Greek).

POPOV, I. 1955. *Proporcii v Bulgarskata Arhitektura (Proportions in Bulgarian Architecture)*. Sofia: Hayka y Yekystvo (in Bulgarian).

RAPPOPORT, P. A. 1995. *Building the Churches of Kievan Russia*. Hampshire: Variorum

SISA, B. 1990. Hungarian Wooden Belltowers of the Carpathian Basin. Pp. 305-352 in *Armos*, vol. A. Thessaloniki: Aristotle University of Thessaloniki, Polytechnic School, Department of Architecture.

STOICHKOV, J. 1977. *Za Arhitecturata na Koprivchtitza (The Architecture of Koprivchtitza)*. Sofia: Tehnika (in Bulgarian).

SUBOTIĆ, G. 1980. *Ohridska Slikarska Škola XV Veka* (L'Ecole de Peinture d'Ohrid au XVe Siecle). Belgrade. (in Serbian).

About the author

Aineias Oikonomou is an architect and for two years was an adjunct lecturer at the Departments of Architecture of University of Patras and Democritus University of Thrace. He earned a Ph.D. and M.Sc. degree in Architecture from the National Technical University of Athens, Greece, and he completed a one-year post-doctoral research in the Laboratory of Architectural Form and Orders, N.T.U.A. School of Architecture. His research interests include architectural morphology, restoration of monuments and historical buildings, study of pre-industrial building techniques, history of construction technology and Islamic architecture.

Farah Habib

Department of Art and Architecture
Science and Research Branch
Islamic Azad University
Tehran, IRAN
f.habib@srbiau.ac.ir

Iraj Etesam

Department of Art and Architecture
Science and Research Branch
Islamic Azad University
Tehran, IRAN
ietessam@hotmail.com

S. Hadi Ghoddusifar*

*Corresponding author
Department of Art and Architecture
Science and Research Branch
Islamic Azad University
Tehran, IRAN
h.ghoddusifar@srbiau.ac.ir

Nahid Mohajeri

Department of Geography,
University College London/

Department of Architecture
Islamic Azad University
South Tehran Branch
Tehran, IRAN
nahid.mohajeri.09@ucl.ac.uk

Research

Correspondence Analysis: A New Method for Analyzing Qualitative Data in Architecture

Abstract. This article aims at establishing a new application of the correspondence analysis (CA) method for analyzing qualitative data in architecture and landscape architecture. This method is primarily used in genealogy but is here, for the first time, applied to architectural studies. After introducing a qualitative method based on coding process, a practical guide for using CAQDAS (Computer-Assisted Qualitative Data Analysis Software) is provided. The software NVivo-8 is applied to analyze the data. CA, a multivariate statistical technique, is used to identify the underlying structure of the data and visualize the results. For the purpose of testing this method in practice, the National Garden of Tehran was selected as a case study to provide the data. The focus is on visualizing the similarities between the properties of the National Garden of Tehran and several different garden design styles. Two reliability tests were performed to verify the results, indicating that the National Garden of Tehran has many characteristics similar to those of a typical Baroque garden style. We believe that this new method may have wide application possibilities for studies on architecture, urban design, and landscape architecture.

Keywords: Landscape architecture, computer technology, statistics, qualitative data analysis (QDA), correspondence analysis (CA)

1 Introduction

Quantitative and qualitative data analyses are both fundamental research methods in architecture and landscape architecture. The types of data primarily used in architecture studies are qualitative, and include texts, photos, maps and drawings. For all these data, qualitative analysis plays an important role. Recent developments in qualitative data analysis methods have opened up new possibilities for social researchers and made significant contributions to studies using empirical data in social sciences.

Coding and classification of the data are among the most common methods in qualitative data analysis. Here "coding" refers to the data being categorized to facilitate analysis, commonly with a view to computer processing. Coding and classification of data are much used in many disciplines within the fields of social and medical sciences. Despite the common use of these methods in social sciences, few scholars in architecture have used coding analysis and computer technology in their studies. Among the few recent architecture studies that have employed coding analysis we may mention Castiglia and Bevilacqua [2008], Eilouti [2008], and Wahba [2010]. More specifically, Wahba [2010] used data classification and comparison methods to analyze data relating to energy-shape and beauty of constructions. Similarly, Castiglia and Bevilacqua [2008] employed synthetic methods to analyze the vaults in Kala's Hammam in El Basan. When it is possible to convert qualitative data into quantitative data, comparison between items

becomes much easier. As an example of this, Eilouti [2008] converted qualitative data, namely the photos of architectural facades, into quantitative codes and then analyzed the façade morphology of several buildings through finite state automaton (FSA).

The principal aim of this paper is to use the correspondence analysis (CA) method to analyze and visualize quantitative architectural data. The case study analyzed here aims to determine which garden design style had the greatest influence on the National Garden of Tehran. The data are derived from comparisons between the National Garden of Tehran and various well-known garden design styles. While some of the present methods have been used in social sciences, as indicated above, they have not previously been applied in architectural studies. Here we employ coding processes using the software NVivo-8 to convert the qualitative data to quantitative data and then to visualize the results using the CA method.

2 Methodologies

2.1 Coding processes

Among the most important methods for qualitative data analysis is the "coding method". Initially, the coding method was introduced by Glaser and Strauss [1965, 1967] in relation to health issues. Subsequently, considerable efforts have been made to improve the coding process and its compatibility with different fields and disciplines. DeCuir-Gunby et al. [2011] discuss the general structure of the coding method and show how to use codebooks (a document type used for gathering and storing codes) in qualitative studies. There have been many discussions in scientific literature about coding methods and codebooks (e.g., [Ryan and Bernard 2000; Mason and Dicks 2001; Franklin et al 2010; DeCuir-Gunby et al 2011]). Nevertheless, experts in qualitative research methods have not yet been able to establish a universally acceptable set of coding procedures [Coffey and Atkinson 1996; DeCuir-Gunby et al. 2011].

Codes can be developed a priori from existing theories or concepts (theory-driven). They can also emerge from the basic data (data-driven) and may also be obtained from a specific project aims and questions (structural codes). Most codes, however, are either theory-driven or data-driven [Ryan and Bernard 2000].

Using a codebook is regarded as an initial and crucial step in the coding process [DeCuir-Gunby et al. 2011]. Housley and Smith [2011] provide the definition that coding should be regarded as a part of analytical procedure by which data, both quantitative and qualitative, are categorized so as to facilitate the analysis process. In fact, the classification of data is a principal step to prepare data for computer processing when using statistical software. The coding method is commonly used to transform the data into a form which is suitable for a computer and thereby produce a theoretical framework for research [Glaser and Strauss 1965, 1967]. Seidel and Kell [1995] suggest three steps in relation to the coding process: 1) consider the events relevant to your study; 2) collect instances of the events; 3) analyze the events so as to find similarities and differences as well as patterns and structures.

To facilitate and increase the validity of research, Computer-Assisted Qualitative Data Analysis Software (CAQDAS) is generally used for the coding process. Initially, CAQDAS emerged in the mid- to late-1980s [Mangaberia, Lee and Fielding 2004]. The rapid development of CAQDAS has been of considerable help as regards importing, storing, grouping, and coding data [Mehmetoglu and Dann 2003]. Mason and Dicks [2001] suggest that theory and the new computer-mediated communication technology coding will converge into a combined, successful method for many types of qualitative data analyses.

MacMillan and Koenig [2004], Alexa and Zuell [2000], and Mangaberia et al. [2004] discuss the basis of the different types of the CAQADS software. Some other studies have addressed the application of CAQADS software to qualitative data analysis (e.g., [Hwang 2008; Pan et al. 2008; Sin 2008; Housley and Smith 2011; Zapata-Sepulveda et al. 2011]. It has been noted by many that CAQDASs are not softwares for automatically analyzing the data; they must rather be regarded as computer tools for helping researcher to facilitate the data analysis [Sin 2008; Weitzman and Miles 1995]. For this reason, there have been many studies intended to demonstrate and guide the applications of such softwares in different fields [Housley and Smith 2011]. There are many different types of the software, but the most commonly used are Atlas.ti, NVivo, and Hyper Research.

NVivo is a popular qualitative data analysis software which provides useful tools for academic research (see www.qsrinternational.com). Zapata-Sepulveda et al. [2011] suggest that NVivo increases the validity and reliability of the results. An important advantage of this software in comparison with others is its ease of use. Also, the software is compatible with different languages and can load data in different formats, including PDF, text, audio, video and image.

There are two different types of code structure in NVivo: "tree node" and "free node". If the data and their associated codes are organized according to content, themes, and research aims in a hierarchical structure, then the method is known as a "tree node" structure [www.qsrinternational.com]. However, if the data is just classified with the aim of finding some patterns or structures that can lead to a general hypothesis, then it is known as a "free node" structure. The tree node structure is more akin to rationalism and the deduction method in science, whereas the free node is more akin to empiricism and the inductive method in science. Tree node structure is used for the coding process in this present research.

2.2 Corresponding analysis (CA) method

Correspondence analysis is an exploratory/descriptive data analysis technique that is used to identify the systematic relations between two or more categorical variables [Hoffman and Frank 1986; Bendixen 1996; Beh 2004; Duller 2010; Greenacre 2007, 2010]. The technique applies primarily to a cross-tabulation or a contingency table of two or more categorical variables, containing discrete or categorical measurements. However, the method can be extended to frequency tables, ratio-scale data in general, binary data, and preferences and fuzzy-coded continuous data [Greenacre 2010: 75]. This means that one can examine the relations not only among row or column variables but also between them.

CA provides useful information about the data and allows one to explore the underlying structure of categorical variables [Greenacre 2007, 2010; Doey and Kurta 2011]. More specifically, the CA technique is a tool to transform a table of numerical information, commonly with a large number of data, into a graphical display known as a biplot (Gabriel biplot), in which each row and each column is shown as a point. A biplot often makes comparison between numerical information easy and helps identify the relations between the different elements under consideration.

A simple scatter plot of two variables has two perpendicular axes, namely the horizontal x-axis and the vertical y-axis. By contrast, a biplot has as many axes as the existing variables, which can have any orientation in the display. Fig. 1 shows the difference between a general scatter plot and a biplot [Greencare 2010: 9].

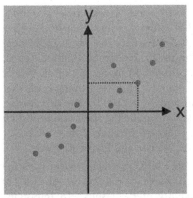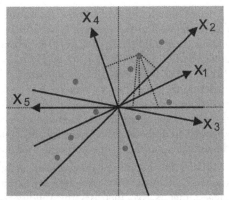

Fig. 1. A simple scatter plot of two variables (left) and a biplot of many variables (right). Red dots represent data points and axes represent variables (see [Greencare, 2007, 2010]).

The main purpose of CA is to reduce the dimensions of the contingency table and visualize the data in a two-dimensional or possibly a three-dimensional graph. One advantage is that the graphical display in CA is in a space of reduced dimensionality, usually two-dimensional, compared to the true dimensionality of the data. The analytical part of the CA method consists then in finding the correct location of points and the orientations of the axes in this reduced space so as to approximate the data as closely as possible. Given that the aim of this paper is primarily to explore applications of the CA method in architecture, the mathematics involved in calculating dimensions is beyond the scope of the paper, for which the reader is referred to standard treatments (e.g, [Clausen 1998; Greencare 2007, 2010]).

There are many concepts used in the CA method. These are defined and discussed in detail in several textbooks [e.g, Clausen 1998; Greenacre 2007, 2010]. Some of the principal concepts may be briefly defined as follows:

- **Margin frequency.** The total frequency of each row or column is named the margin frequency of that row or column. The concept is very useful when calculating row and column profiles and masses.

- **Grand total.** This refers to the total margin frequency of all rows or all columns.

- **Row and column profile.** The row and column profile is equal to the frequency of each element divided by the margin frequency of that row or column.

- **Row and column mass.** In the CA method, the mass of each column/row is equal to the margin frequency of that column/row divided by the grand total frequency of the elements.

- **Centroid.** In the CA method, the centroid is the geometric center of the graph. The row and column profiles and mass have a great effect on centroid of the graph.

- **Inertia** (here meaning the moment of inertia). In the CA method, the total inertia value, also known as variance, describes the level of association, or dependence, between variables. It shows how well the row and column profile points are represented in the graphical display.

- **Chi-square (χ^2) statistic** (a weighted Euclidean distance). This is a measure of the discrepancy between the observed and expected frequencies. CA uses the chi-

square statistic to measure the distance between points on the biplot [Ebdon 1985; Clausen 1998; Greencare 2010; Navidi 2011]. In addition, the chi-square distance measures the association between variables. The chi-square distance can be used to examine the relation between categories of the same variables but not the between variables of different categories.

In this paper, much use is made of the concept of a contingency table N (I, J). The table displays the frequency for each combination of two or more variables, that is, n_{ij}, with I rows (i=1, 2, ... I) and J columns (j =1,2,..., J). To clarify the terms related to the contingency table, the following definitions may be helpful. Row and column margin frequencies are denoted by n_i+ and n+j respectively and are given by

$$n_{i+} = \sum_j n_{ij} \tag{1}$$

and

$$n_{+j} = \sum_i n_{ij} \tag{2}$$

The total margin frequencies or the grand total are given by

$$n = \sum_j \sum_i n_{ij} \tag{3}$$

The r_{ij} and c_{ij} are row and column profiles are given respectively by

$$r_{ij} = \frac{n_{ij}}{n_{i+}} \tag{4}$$

$$c_{ij} = \frac{n_{ij}}{n_{+j}} \tag{5}$$

where r_i and c_j are the mass of row i and the mass of column j, respectively, and defined as

$$r_i = \frac{n_{i+}}{n} \tag{6}$$

$$c_j = \frac{n_{+j}}{n} \tag{7}$$

Each row and column profiles of the table have an inertia which is a portion of the total inertia. The inertia of ith row profile is given by

$$Inertia = r_i \sum_j \frac{(r_{ij} - c_j)^2}{c_j} \tag{8}$$

The inertia of jth column profile is computed similarly. Accordingly, the total inertia of the contingency table is given by the square of "phi-coefficient", namely as

$$\phi^2 = \sum_i \sum_j \frac{(p_{ij} - r_i c_j)^2}{r_i c_j} \tag{9}$$

Here p_{ij} is relative frequency and is given by

$$p_{ij} = \frac{n_{ij}}{n} \tag{10}$$

Equation (9) can also be written as

$$\phi^2 = \frac{\chi^2}{n} \tag{11}$$

which is Pearson's chi-square statistic divided by the grand total n. The chi-square statistic is then given by

$$\chi^2 = \sum \frac{d^2}{e} \tag{12}$$

where χ^2 is the symbol for chi-square, d is the difference between the observed and the expected frequency for each variable in each category, and e is the associated expected frequency. Each expected frequency can be calculated by:

(column margin frequency × row margin frequency) / grand total.

3 Application

3.1 The National Garden of Tehran (Baq-e-Melli)

The National Garden of Tehran is located in the city of Tehran, has been chosen as the case study for applying the present method. The National Garden of Tehran was built in 1928 (in the first Pahlavi period of the Persian calendar). The place was the first public green space built at that time, and is recognized as the first city park in Iran. This square-shaped park was built as a part of a large city square but was destroyed around five years later in 1933 (fig. 2). The basic data from the garden used in this study include aerial photographs, texts, newspaper images, magazine documents, and images from private archives.

3.2 Coding method

Several sources (e.g., [Turner 2005; Rohani 2010; Shahcheragi 2010]) were studied to collect the relevant information about the historical background of different garden design styles worldwide. These were used to create the associated codes in connection with the National Garden of Tehran. From the data collected, the main points are marked with a series of codes. The identified codes, in addition, reflect the aims of the study.

The garden styles selected were those regarded as most likely to have influenced the design of the National Garden of Tehran. The properties of the chosen garden styles were identified and classified in view of the research aims. A total of seventy-two sub-codes were marked based on different properties of garden design styles and the National Garden of Tehran. Then, all seventy-two sub-codes were grouped into seven main groups of codes. These seven selected main groups of codes are related to five main garden design styles that were thought most likely to have influenced the design of National Garden of Tehran. We choose two coders to do the tests of research reliability. The details are discussed below in the section on reliability tests.

Fig. 2. The National Garden of Tehran. a) An aerial view. The garden, marked with dotted-red square, is located in the west part of the Mashgh square. The entrance is in east, marked with a red arrow. Source: [Khadem 1928]; b) A view of the entrance. Source: [Tahami 1928]; c) Sketch map of the layout of the park based on the aerial photo. Source: the authors.

The seven main groups of codes, all of which refer to the properties of the National Garden of Tehran (Table 1), are 1) *decoration*; 2) *geometry and axes*; 3) *nature and its elements*; 4) *order*; 5) *architecture and artistic elements*; 6) *relation with city and environment*; 7) *function*. Some main codes (properties) have sub-codes. For example, *decoration* has sub-codes that refer to all decorative figures used in gardens, including fountains, basins, garden pavings, gravel paths, and ornamental flower as well as trees or bushes trimmed into ornamental shapes. Similarly, *geometry and axes* has sub-codes that refer to organic and grid patterns of pathways, symmetric and asymmetric axial views and perspectives, as well as regular and irregular shapes of parterres. Also, *nature and its elements* has sub-codes that refer to all kind of natural elements used in gardens, including water, types of trees, plants, and evergreen shrubs. So does architecture and artistic elements, where the sub-codes refer to all architectural structures inside the gardens, such as entrance, walls, stairways, fences, statues and sculptures, as well as pavilions. The sub-codes of *relation to city and environment* refer to the extent to which the gardens were separated from their surroundings and whether they could be characterized as "introverted" or "extroverted". However, two main codes do not have sub-codes. These are *order*, which refers to general layouts of gardens and *function*, which reflects the degree to which the gardens were used as private versus public places.

The numbers in Table 1 indicate the frequencies of common properties, that is, how many sub-codes are common between the garden design styles and the properties of the

National Garden of Tehran. For example, 0 means that there are no common sub-codes, 1 means there is one common sub-code and so on. Similarly, 6 indicates that there are six sub-codes under the property of the *geometry and axes* which the National Garden of Tehran has in common with the Baroque style. Thus, the higher the number, the greater is the similarity between the garden design styles and the National Garden of Tehran. As mentioned above, two main codes, namely *function* and *order*, do not include any sub-codes. As a consequence, the highest value in the corresponding columns is 1 (Table 1). For example, Persian gardens (row 3 in Table 1) are mainly private gardens: therefore, they have no similarity with the National Garden of Tehran and get the code number of 0. However, the National Garden of Tehran is a public garden, similar to the Industrial Revolution Public Park, and gets the code number of 1.

3.3 Coding process using CAQDAS

Here NVivo-8 is used for the coding processes of the National Garden of Tehran. All the data, including the photos, drawings and texts, were loaded into the software. Then the previously established and defined codes were imported. As mentioned before, in this research tree node structure is used for coding processes. Whenever there was similarity between the properties of the National Garden of Tehran and the identified garden design styles, the related code was allocated to the part of the data indicating that similarity. Fig. 3 shows a representation of coding process in NVivo-8.

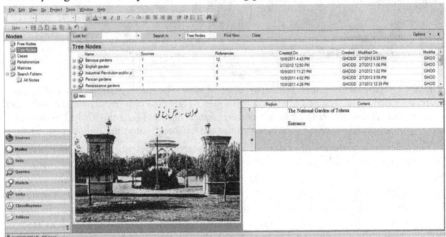

Fig. 3. Coding process of the National Garden of Tehran by NVivo-8

3.4 Visualizing quantitative data

The next step after the coding process is to prepare the quantitative data in a suitable format, and make them ready for visualization. Statistical plots are one of the most important forms of quantitative data analysis, and very useful tools for visualization of data. Thus, visualizing the data through histograms and bar charts helps to understand better and interpret accurately the relations between different variables [Pan, Chon and Song 2008; Greenacre 2010: 9].

Although NVivo has tools for visualizing several types of graphs, its capabilities for visualization are quite limited. Thus, using different statistical softwares can be appropriate and helpful (e.g., [Pan et al. 2008; Zapata-Sepulveda et al. 2011]). For example, after coding process by NVivo the results can be exported to Excel or SPSS applications for manipulation of the quantitative data [Sin 2008] and make them ready

for visualization. Table 1 shows the results of the prepared data in relation to garden design styles and the properties of the National Garden of Tehran. As mentioned above, in the coding method, the frequencies show the degree of similarity between the garden design styles and the National Garden of Tehran. As mentioned before, the lowest value, namely 0, indicates no similarity between the corresponding garden design style and the associated property of the National Garden of Tehran. For example, the English style and the National Garden of Tehran have no similarity as regards *decoration*. In contrast, as regards *geometry and axes* properties, the highest degree of similarity is between the Baroque style and the National Garden of Tehran.

Garden styles	The Properties of the National Garden of Tehran						
	Decoration	Geometry and Axes	Nature and its Elements	Order	Architectural and Artistic Elements	Relation with City and Environment	Function
Baroque	3	6	1	1	1	0	0
English	0	1	1	0	2	0	0
Persian	0	2	1	0	2	1	0
Renaissance	2	2	1	0	0	2	0
Industrial Revolution Public Park	0	4	0	1	0	3	1

Table 1. Correspondence table shows the frequencies of each variable in each category. There are two categories, namely garden design styles, with 5 variables, and properties of National Garden of Tehran, with 7 variables

To visualize the results in a statistical graph, the cumulative frequency of each property in relation to garden design styles were calculated (each row). According to fig. 4, the Baroque garden was the most influential garden style in the National Garden of Tehran with a total consistency degree of 12.

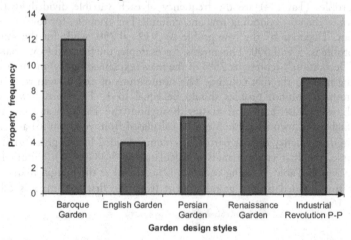

Fig. 4. The frequency plot of the National Garden of Tehran properties and garden styles

However, column graphs appear to be somehow limited in their ability to display all the data in large contingency tables [Greenacre 2010: 9]. To deal with the limitations of column graphs, in the case of having large number of data, a new and more accurate tool can be used to visualize the data. CA is one of the most common statistical graphs which can be used for displaying information clearly and in an easily understandable graphical diagram.

3.5 Correspondence analysis of the National Garden of Tehran

In this study, CA is used to visualize the similarities between the garden styles and the properties of the National Garden of Tehran. This method has been used in many fields, but is here for the first time applied to genealogy studies in architecture and landscape architecture. This section describes the application of the CA method to the chosen case study, the National Garden of Tehran.

Using this method has several advantages over other techniques of data analysis, namely the following:

1. CA, as a multivariate method, can be used to visualize two or more categorical variables. Therefore, it can reveal relations between the data which would not be detected in the case of pairwise or comparison variables;

2. The results of the CA method are shown in graphical display, known as a biplot, of rows and column points and can be used to detect the structural relations among the categorical variables;

3. The data format for using the CA technique is very flexible, although the data has to be in matrix format with a non-negative value.

Table 1 shows the frequencies of all variables, obtained from the coding process. The rows correspond to garden design styles and the columns correspond to the properties of National Garden of Tehran. The margin frequencies for rows and columns can be calculated using equations (1, 2), as shown in Table 2. The grand total or total margin frequency, calculated from equation (3), is equal to 38 (Table 2). This table has been used as a base to calculate the relative frequencies, known as row profiles and column profiles, across all cells using equations (4) and (5). The row profiles (Table 3) and column profiles (Table 4) are the frequency of each variable divided by the margin frequencies of the corresponding row and column. For example, for row "Baroque" and the column "Decoration" the row profile is: 3/12= 0.250, while for the same cell the column profile is: 3/5=0.600. This means, for example, the first cell (row "Baroque" and column "Decoration") represents 25% of the mass (explained below) in the first row and 60% of the mass in the first column. The summations of each row in row profiles and each column in column profiles should be equal to 1. Therefore, row and column profiles show the distribution of masses throughout the Tables 3 and 4. The column masses which are shown in Table 3 can be calculated from equation (6) and are defined as the margin frequency of that particular column divided by the grand total n ($n=38$). For example, the first column mass is calculated as 5/38=0.132). Similarly the row masses (shown in Table 4), using equation (7), defined as the margin frequency of that particular row divided by the grand total n (e.g., the first row mass is calculated as 12/38=0.316).

The Properties of the National Garden of Tehran

Garden styles	Decoration	Geometry and Axes	Nature and its Elements	Order	Architectural and Artistic Elements	Relation with City and Environment	Function	Margin Frequency
Baroque	3	6	1	1	1	0	0	12
English	0	1	1	0	2	0	0	4
Persian	0	2	1	0	2	1	0	6
Renaissance	2	2	1	0	0	2	0	7
Industrial Revolution Public Park	0	4	0	1	0	3	1	9
Margin Frequency	5	15	4	2	5	6	1	38

Table 2. Correspondence table for the margin frequencies of each row and column and the grand total

The Properties of the National Garden of Tehran

Garden styles	Decoration	Geometry and Axes	Nature and its Elements	Order	Architectural and Artistic Elements	Relation with City and Environment	Function	Total
Baroque	0.250	0.500	0.083	0.083	0.083	0.000	0.000	1.000
English	0.000	0.250	0.250	0.000	0.500	0.000	0.000	1.000
Persian	0.000	0.333	0.167	0.000	0.333	0.167	0.000	1.000
Renaissance	0.286	0.286	0.143	0.000	0.000	0.286	0.000	1.000
Industrial Revolution Public Park	0.000	0.444	0.000	0.111	0.000	0.333	0.111	1.000
Column Mass	0.132	0.395	0.105	0.053	0.132	0.158	0.026	1.000

Table 3. Row profiles and the column-mass values across the bottom

Garden styles	Decoration	Geometry and Axes	Nature and its Elements	Order	Architectural and Artistic Elements	Relation with City and Environment	Function	Row Mass
Baroque	0.600	0.400	0.250	0.500	0.200	0.000	0.000	0.316
English	0.000	0.067	0.250	0.000	0.400	0.000	0.000	0.105
Persian	0.000	0.133	0.250	0.000	0.400	0.167	0.000	0.158
Renaissance	0.400	0.133	0.250	0.000	0.000	0.333	0.000	0.184
Industrial Revolution Public Park	0.000	0.267	0.000	0.500	0.000	0.500	1.000	0.237
Total	1.000	1.000	1.000	1.000	1.000	1.000	1.000	1.000

Table 4. Column profiles and the row-mass values

Each row profile of Table 3 and column profile of Table 4 has an inertia which is a fraction of the total inertia (indicated in Tables 5 and 6), which can be calculated from equation (8) (indicated in Tables 7 and 8). For example, the following example shows the calculated inertia for the first row, "Baroque", which appears in the overview of row points in Table 7.

$$= 0.316 \left[\frac{(0.25 - 0.132)^2}{0.132} + \frac{(0.500 - 0.395)^2}{0.395} + \frac{(0.083 - 0.105)^2}{0.105} + \frac{(0.083 - 0.053)^2}{0.053} + \frac{(0.083 - 0.132)^2}{0.132} + \frac{(0.000 - 0.158)^2}{0.158} + \frac{(0.000 - 0.026)^2}{0.026} \right]$$

$$= .316 \left[0.105 + 0.028 + 0.005 + 0.017 + 0.018 + 0.158 + 0.026 \right]$$

$$= 0.316 \times 0.357$$

$$= 0.113$$

The total inertia is simply calculated from equations (9) and (10) and is equal to 0.691 (Table 5). The total inertia or variance in CA has a close connection with the chi-square statistic and is defined as a measure of statistical association between rows and columns. The total inertia can also be calculated from equation (11), that is, the chi-square statistic (26.248) divided by the grand total (38). The chi-square statistics for rows and columns, as well as the total chi-square statistic, are calculated from equation (12) and shown in Table 6. The following example shows the calculated chi-square statistics for the first row, namely Baroque (marked in bold), using equation (12) and the simple rule for calculating expected frequency cited earlier, that is, column margin frequency times row margin frequency divided by grand total (the calculated expected frequencies are not presented here):

$$\chi^2 = \frac{(3 - 1.579)^2}{1.58} + \frac{(6 - 4.737)^2}{4.74} + .\frac{(1 - 1.263)^2}{1.26} + \frac{(1 - 0.632)^2}{0.63} + \frac{(1 - 1.579)^2}{1.58} + \frac{(0 - 1.895)^2}{1.89} + \frac{(0 - 0.316)^2}{0.32}$$

$$= 1.279 + 0.337 + 0.055 + 0.214 + 0.212 + 1.895 + 0.316$$

$$= 4.308$$

| | | | | | Proportion of Inertia | | Confidence Singular Value | |
| | | | | | | | | Correlation |
Dimension	Singular Value	Inertia	Chi Square	Significant (p-value)	Accounted for	Cumulative	Standard Deviation	2
1	0.607	0.368	–	–	0.533	0.533	0.106	0.354
2	0.471	0.222	–	–	0.322	0.855	0.095	
3	0.309	0.096	–	–	0.138	0.993		
4	0.070	0.005	–	–	0.007	1		
Total		0.691	**26.248**	0.341	1	1		

Table 5. Summary of the main statistical results

The Properties of the National Garden of Tehran

Garden styles	Decoration	Geometry and Axes	Nature and its Elements	Order	Architectural and Artistic Elements	Relation with City and Environment	Function	Total
Baroque	1.279	0.337	0.055	0.215	0.212	1.895	0.316	**4.308**
English	0.526	0.212	0.796	0.211	**4.126**	0.632	0.105	6.608
Persian	0.789	0.057	0.215	0.316	1.856	0.003	0.158	3.394
Renaissance	1.264	0.211	0.094	0.368	0.921	0.724	0.184	3.767
Industrial Revolution Public Park	1.184	0.056	0.947	0.585	1.184	1.754	2.459	8.170
Total	5.043	0.874	2.107	1.694	8.300	5.008	3.222	**26.248**

Table 6. Chi-square statistics for rows and columns and the total

The statistical summary shows a variety of useful information (Table 5). The maximum number of dimensions that can be extracted from a two-way contingency table is equal to the minimum of the number of columns minus 1, and the number of rows minus 1. Therefore, 4 dimensions were extracted. If we choose to extract the maximum number of dimensions, then we can represent exactly all the information contained in the table. By reducing the dimensions of the graph, some information will be lost. But the amount of information that can be shown in a two-dimensional CA graph is related to the proportion of inertia for each dimension, which is a high proportion for the first two dimensions. Generally, in a CA graph, two dimensions are used for showing the results, and therefore that percentage of total variability that is explained by the first two dimensions.

A singular value is simply the square root of an inertia value, which describes the maximum linear correlation between the categories of the variables for any given dimension. Singular values can be interpreted as correlations between the row and column scores in dimension shown in Tables 7 and 8. For example, the singular value for dimension 1 is 0.607, that is, the correlation between the row and column scores for dimension 1. Similarly, the singular value for dimension 2 is 0.471, that is, the correlation between the row and column scores for dimension 2.

The total inertia value represents the variance of the data in the correspondence table. The inertia value for each dimension, therefore, reflects the relative importance of that dimension. The first dimension (0.607) and the second dimension (0.471) are always the most important ones.

The chi-square statistic tests the hypothesis that the total inertia value is/is not different from zero by simply looking at the p-value. The p-value measures the plausibility of hypothesis. The p-value here is greater than 0.05 (a common cut-off value), which indicates that the total inertia is significantly different from zero, and therefore that the hypothesis is plausible. It is important to note that CA is a non-parametric statistic. This means there is no theoretical distribution to which the observed distances can be compared. Therefore, contrary to the classical applications of the chi-square test, using chi-square test in CA does not reveal whether the association between variables is statistically significant. In fact, CA uses the chi-square statistic only to test the plausibility of total inertia value.

The proportion of inertia columns represents the contribution of total inertia for each dimension. It is calculated as each inertia value divided by the total (0.368/0.691=0.533). For example, the first dimension (0.368) explains 53.3% of the total inertia (0.691). However, dimension 1 and 2 together (0.59) explain 85% of the total inertia, that is, the variation in the data. The standard deviation column refers to the standard deviation of the singular values and shows the relative precision of each dimension. Finally, the correlation column refers to the correlation between dimensions.

To visualise the row and column points and their relations in the correspondence map in two dimensions, 1 and 2, the row and column coordinates are calculated, using the statistical program SPSS, and the results shown in Tables 7 and 8. The overview row points (Table 7) shows masses, scores in dimension (coordinate points), inertia, the standard deviations of each row score in dimensions 1 and 2, and finally their correlations. The contribution of each row to the dimensions and each dimension to the rows were also calculated (Table 7). The mass for each row is already explained (Table 4). The score in dimensions shows each row's score on dimension 1 and dimension 2. The scores are derived based on the proportions (mass) for each cell, column and row, when compared to the grand total n. The scores are representative of dimensional distance and are used as coordinates for points when plotting the correspondence map (CA graph) below (fig. 5). The inertia column shows the amount of variance of each row explained by the total inertia value. The confidence row points show the standard deviations of the row scores (the values used as coordinates to plot the correspondence map) and are used to assess the score precision, as well as the correlation between the row dimension scores. The overview column points (Table 8) are similar to the row points.

Garden styles	Mass	Score in Dimension		Inertia	Confidence Row Points		
					Standard Deviation in Dimension		Correlation
		1	2		1	2	1-2
Baroque	0.316	0.007	0.758	0.113	0.658	0.446	-0.229
English	0.105	1.573	-0.527	0.174	0.692	0.897	0.369
Persian	0.158	0.795	-0.560	0.089	0.714	0.525	0.121
Renaissance	0.184	-0.339	0.535	0.099	0.522	0.919	0.199
Industrial Revolution Public Park	0.237	-0.975	-0.819	0.215	0.540	0.527	-0.884
Total	1.000			0.691			

Table 7. Overview of row points

Property	Mass	Score in Dimension		Inertia	Confidence Row Points		
					Standard Deviation in Dimension		Correlation
		1	2		1	2	1-2
Decoration	0.312	-0.216	1.418	0.133	0.937	0.300	0.095
Geometry and Axes	0.395	-0.151	0.098	0.023	0.332	0.401	0.248
Nature and its Elements	0.105	0.839	0.109	0.055	0.369	0.601	0.023
Order	0.053	-0.798	-0.065	0.045	0.605	1.029	0.469
Architectural and Artistic Elements	0.0132	1.564	-.600	0.218	0.496	0.727	0.840
Relation with City and Environment	0.158	-0.771	-.688	0.132	0.648	0.846	-0.142
Function	0.026	-1.607	-1.737	0.085	1.714	1.313	0.039
Total	1.000			0.691			

Table 8. Overview of column points

To show the results, first we plot the row and column points separately (Tables 7 and 8). The two graphs in fig. 5 show the column points for National Garden of Tehran properties and row points for garden design styles. The graphs display the scores for each category on both dimensions (dimension 1 horizontal axes and dimension 2 vertical axes) (Tables 7, 8). It is more common to show the row and column coordinates in single plot, called a "biplot" (fig. 6). The biplot correspondence map shows each category scores on both dimensions for both garden design styles and properties of the National Garden of Tehran. The biplot for row and column points can be interpreted in different ways depending on different methods used to standardize the row and column scores given in Tables 7 and 8. Symmetrical normalization, a default method in SPSS, is a technique used here to standardize the row and column scores in order to be able to make general comparison between the data. Symmetrical normalization divides the inertia equally over both the rows and columns. Distances between categories for a single variable cannot be interpreted, but distances between the categories for different variables are meaningful.

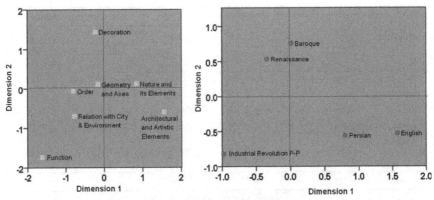

Fig 5. Separate plots for the rows and columns points

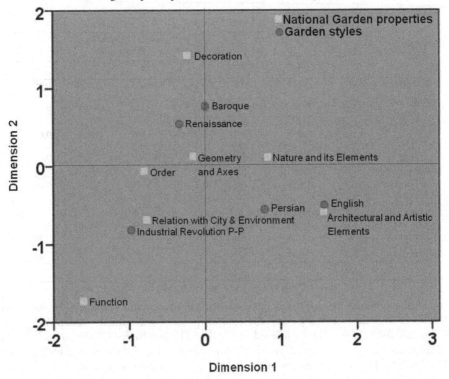

Fig. 6. Biplot correspondence map

4 Reliability tests

To explore the results, two reliability tests are used. Reliability tests are different from validity tests. The latter is more common in quantitative research, whereas the former is used more in qualitative research, namely studies where the data is not based on measurements. According to Moret et al. [2007], reliability and validity tests are both context bound, that is, depend much on the context in which the tests are made. Moret et al. [2007], Krippendroff [2004], Franklin et al. [2010] and Lombard et al. [2002] discuss the advantages of reliability tests in qualitative research. Krippendroff [2004] has introduced seven methods to assess the reliability based on different types of calculations.

The most common reliability methods include "Percent agreement", "Osgood's CR", "Bennett's S", "Scott's π", "Krippendorff's α", "Cohen's κ", and "Benini's β". In the present research we used two specific methods, namely Scott's π (Scott's pi) and Cohen's κ (Cohen's kappa coefficient) for estimating the reliability of the results. The two methods are similar and assess the extent of agreement between the coders, except that the expected agreement is calculated slightly differently in these two methods. More specifically, Cohen's κ is a measure of how well the coders or raters (the persons who rate the similarities between items) agree among themselves, that is, a measure of the inter-annotator agreement [Krippendroff 2004; Strijbos et al. 2006]. In this case, the two persons are the coders number 1 and 2, such as those coders that rate the similarities in Table 9. It should be clear that the reliability tests are made to compare the ratings (made by the two coders) of the similarities between the garden styles and the properties of the National Garden of Tehran. This was done to verify the results in Table 1. This is necessary because the ratings in Table 1 were made by a computer, through the coding process program NVivo-8.

The other method, Scott's π, is also used to assess how well the raters agree among themselves. However, while Cohen's κ makes no assumptions as to the distribution of responses (answers or assessments by the coders), Scott's π assumes that the distributions of responses or answers from the coders are the same [Cramer, 1994]. Tables 9 and 10 show the coding results. Considering the Cohen's κ and Scott's π values, the results of the research are reliable. The detail discussions of the two methods are as follows.

		Coder 1		
		Coded data (A)	None-coded data (B)	Total (C)
Coder 2	Coded Data (1)	35	4	39
	Non-coded data (2)	2	31	33
	Total (3)	37	35	72

Table 9. Cohen's κ

Cohen's κ is obtained from the following equation:

$$\kappa = \frac{P - P_e}{1 - P_e} \tag{13}$$

where P is the "observed proportion of agreement" and calculated based on the number of items on which coders agreed divided by the total number of items: that is, the sum of the data in A1 (35 data) and B2 (31 data) divided by C3 (72), or a total of 0.917. The "chance expected proportion of agreement or P_e" is obtained from the following equation, with symbols as in Table 9:

$$P_e = \frac{C1}{C3} \times \frac{A3}{C3} + \frac{C2}{C3} \times \frac{B3}{C3} \tag{14}$$

From Eq. (14) we find that P_e is equal to 0.5. With reference to Table 9, we obtain the Cohen's κ equal to 0.83. The maximum value for κ is when the observed level of agreement is κ = 1, and generally κ ≤ 1. A value of 1 implies a perfect agreement and

values less than 1 imply less than perfect agreement. It is difficult to answer the question how large κ should be to indicate a good agreement; however, the most common level selected is κ > 0.8 [Krippendroff, 2004, Cramer, 1994]. The κ value (0.83) here is comparatively high; it means that the results are reliable. We also calculate the Scott's π to compare the results with the Cohen's κ. Table 10 shows the details of Scott's π reliability test.

Join Proportion Squared (D)	Join Proportion (C)	Coder 2 (B)	Coder 1 (A)	
0.27	0.52	39	37	Coded Data (1)
0.22	0.47	33	35	None-coded Data (2)
0.49	----	72	72	Total (3)

Table 10. Scott's π

Scott's π is also obtained from similar equation except P_e in Scott's π is calculated differently than when using Cohen's κ, and is based on the joint proportion. P is calculated based on data in Table 9 (symbols as in Table 9), and is the sum of the data in A1 (35 data) and B2 (32 data) divided by C3 (72) or a total of 0.917. The P_e is simply the sum of the squared values of the join proportion, namely D1 (0.27) and D2 (0.22), or a total of 0.49. The joint proportion for coded data (symbols as in Table 10) is the sum of the data in A1 (37) and B1 (39) divided by the twice the total of data in A3/B3 (72) or a total of 0.52. The joint proportion for non-coded data is the sum of the data in A2 (35) and B2 (33) divided by the twice the total of data in A3/B3 (72), or a total of 0.47. The calculated Scott's π value, 0.83, is higher than the threshold value of π > 0.80, which means that the results are reliable.

5 Discussion

The results from fig. 3 show that the National Garden of Tehran was more influenced by the Baroque garden style than by other styles. These conclusions can also be seen from the correspondence map (fig. 6). Several properties of National Garden of Tehran, such as geometry and axes, decoration, order, and nature and its elements, are close to the centroid, and therefore to the Baroque garden style. While some generalizations can be made about the association between categories, the correspondence map does not explain the associations between columns and row points.

The results of CA represent the relative closeness of the points (figs. 5, 6). The locations of data points in relation to the centroid show approximately how far they are from the row and column profiles. However, to judge properly the relative closeness of the points, one should look at the horizontal or vertical distance between the points, and not at the oblique (diagonal) distances. In such plots, one can only interpret the distances between row points and the distances between column points but not the distances between row points and column points. However, it is legitimate to interpret the relative positions of one point of one category with respect to all points of the other category.

In a correspondence map the axes on which the plot is constructed are called dimension 1 and dimension 2 (Tables 7, 8). These are not based on observed values, but derived from the data with the aim of explaining the most variance possible. The first

dimension explains 53.3% of the total inertia/variance (Table 5). The space defined by the first two dimensions gives the best solution, in this example, explaining 53.3 + 32.2 = 85.5% of the total variance. Each dimension simply builds on the previous ones (cumulative) to explain additional variance (Table 5), but in decreasing order. A point makes a large contribution to the inertia in two ways: 1) when its distance from the centroid is large (even if it may have a small mass); 2) when it has a large mass (even if the distance is small) [Bendixen, 1996; Beh, 2004; Duller, 2010].

Based on these considerations, the interpretation of correspondence map is as follows. Table 2 and fig. 6 present the influence of several garden design styles on the design of National Garden of Tehran. This shows that the National Garden of Tehran had the greatest similarity to the Industrial Revolution public park as regards function and relationship with the city and environment. However from the point of view of geometry and axes, order, nature and its elements and decoration, it was closer to the Baroque garden style. In addition, considering the architectural and artistic element, the National Garden of Tehran was quite similar to the Persian garden and the Baroque garden.

One main technique of presentation, used here, is the biplot. When using a biplot, the chi-square statistic reveals the strengths of correlation between the data, which is based on the point distances of categories. The distance between any row points or column points in the biplot (Table 6, fig. 6) gives a measure of their similarity (or dissimilarity). Points that are mapped close to one another have similar profiles, whereas points mapped far away from one another have very different profiles. Therefore, the chi-square distances (Table 6) show the contribution of each cell to the overall chi-square statistic, which is a measure of how different the row and column profiles are. In this case, the largest distance comes from the row "English" and the column "Architectural and artistic elements" (the bold value in Table 6). The calculated inertia for all variables in Table 6 also shows the contribution of each cell to the overall variability (details not shown here). In this case, row "English"and column "Architectural and artistic elements" has the highest inertia.

In this paper, Scott's π and Cohen's κ are both used for measuring the reliability of the research. Both calculated values are 0.83. These values should be compared with threshold values. The calculated values, 0.83, are higher than the threshold value of 0.80, which means that the results are reliable.

6 Conclusion

In conclusion, we have introduced the CA method for analyzing qualitative data in architecture and landscape architecture. The original data for determining which garden design style had the greatest influence on the National Garden of Tehran are primarily text, graphs, maps and drawings. The coding method and NVivo-8 software are used to make the data measurable, and convert the qualitative data into quantitative format. In addition, the CA method is used for not only statistical analysis, but also to visualize the association between the rows and columns of the contingency table. Finally, two reliability tests, Cohen's κ and Scott's π, show that the results of the coding process and also that of the CA method are reliable. Although the CA method is applied here only for a genealogical study of a park, it can also be used for other purposes such as for tracing building styles and urban forms, topics that may be regarded as a future development of the present work.

Acknowledgments

We thank the *Nexus Network Journal* reviewer for very helpful comments that greatly improved the paper.

References

ALEXA, Melina and Cornelia ZUELL. 2000. Text Analysis Software: Commonalities, Differences and Limitations: The Results of a Review. *Quality and Quantity* **34**, 3: 299-321.

BEH, Eric J. 2004. Simple Correspondence Analysis: A Bibliographic Review. *International Statistical Review* **72**, 2: 257-284.

BENDIXEN, Mike. 1996. A Practical Guide to the Use of Correspondence Analysis in Marketing Research. *Research on-line* **1**, 16-38.

CASTIGLIA, Roberto B.F. and Marco Giorgo BEVILACQUA. 2008. The Turkish Baths in Elbasan: Architecture, Geometry and Wellbeing. *Nexus Network Journal* **10**, 2: 307-322.

CLAUSEN, Sten Erik. 1998. *Applied Correspondencs Analysis: An Introduction.* Thousand Oaks, CA: Sage.

COFFEY, Amanada Jane and Paul Anthony ATKINSON. 1996. *Making Sense of Qualitative Data: Complementry Research Strategies.* London: Sage.

CRAMER, Duncan. 1994. *Introducing statistics for social research: Step-by-step calculations and computer techniques using SPSS.* London: Routledge.

DECUIR-GUNBY, Jessica T., Patricia L. MARSHALL and Allison W. MCCULLOCH. 2011. Developing and Using a Codebook for the Analysis of Interview Data: An Example from a Prefessional Development Research Project. *Field Methods* **23**, 2: 136-155.

DOEY, Laura. and Jessica KURTA. 2011. Correspondence analysis applied to psycological research. *Tutorials in Quantitative Methods for Pyschology* **7**, 1: 5-14

DULLER, Christine. 2010. Correspondence Analysis – Theory and Application in Management Accounting Research. ICNAAM 2010: International Conference of Numerical Analysis and Applied Mathematics 2010. *AIP Conference Proceedings* **1281**: 1905-1908.

EBDON, David. 1985. *Statistics in Geography.* Cambridge, MA: Basil Blackwell Ltd.

EILOUTI, Buthayna H. 2008. A Formal Language for Palladian Palazzo Facades Represented by String Recognition Device. *Nexus Network Journal* **10**, 2: 245-268.

FRANKLIN, Cynthia S., Patricia A. CODY and Michelle BALLAN. 2010. Reliability and Validity in Qualitative Research. In *Handbook of Social Work Research*, B. A. Thyer, ed. London: Sage.

GLASER, Barney and Anselm STRAUSS. 1965. *Awareness of Dying.* New York: Aldine De Gruyter.

———. 1967. *The Discovery of Grounded Theory: Strategies for Qualitative Research.* Chicago: Aldine.

GREENACRE, Michael. 2007. *Correspondence Analysis in Practice*, 2nd ed. London: Chapman and Hall.

———. 2010. *Biplots in Practice.* Spain: Fundacion BBVA. Available at: http://www.multivariatestatistics.org/biplots.html (last accessed, 28 August 2012).

HOFFMAN, Donna L. and George R. FRANK. 1986. Correspondence Analysis: Graphical Representation of Categorical Data in Marketing Research. *Journal of Marketing Research* **23**, 3: 213-227

HOUSLEY, William and Robin James SMITH. 2011. Telling the CAQDAS code: Membership categorization and the accomplishment of 'coding rules' in research team talk. *Discourse Studies* **13**, 4:417-434.

HWANG, Sangsoo. 2008. Utilizing Qualitative Data Analysis Software: A Review of Atlas.ti. *Social Science Computer Review* **26**, 4: 519-527.

KHADEM, Ali. 1928. Arial Photo of the National Garden, Private Archive. Tehran: Iran

KRIPPENDROFF, Klaus. 2004. Reliability in Content Analysis: Some Common Misconceptions and Recommendations. *Human Communication Research* **30**, 3: 411-433.

LOMBARD, Matthew Jenifer SNYDER-DUCH, and Cheryl Campanella BRACKEN. 2002. Content Analysis in Mass Communication: Assessment and Reporting of Intercoder Reliabiliy. *Human Communication Research* **28**, 4:587-604

MACMILLAN, Katie and Thomas KOENIG. 2004. The Wow Factor: Preconception and Expectations for Data Analysis Software in Qualitative Research. *Social Science Computer Review* **22**, 2: 179-186.

MANGABERIA, Wilma C., Raymond M. LEE and Nigel G. FIELDING. 2004. Computer and Qualitative Reseach: Adoption, Use and Representation. *Social Science Computer Review* **22**, 2: 167-178.

MASON, Bruce and Bella DICKS. 2001. Going Beyond the Code: The Production of Hypermedia Ethnography. *Social Science Computer Review* 19, 4: 445-457.

MEHMETOGLU, Mehmet and Graham M. S. DANN. 2003. Atlas/ti and Content: Semiotic Analysis in Tourism Research. *Tourism Analysis* 8, 1: 1-13.

MEYER, Carl D. 2000. *Matrix Analysis and Applied Linear Algebra.* Philadelphia:Siam.

MILES, Matthew B. and A. Michael HUBERMAN. 1994. *Qualitative Data Analysis: An Expanded Sourcebook.* London: SAGE.

MORET, Margriet, Rob REUZEL, Gert Jan VAN DER WILT and John GRIN. 2007. Validity and Reliability of Qualitative Data Analysis: Interobserver Agreement in Reconstructing Interpretative Frames. *Field Methods* 19, 1: 24-39.

NAVIDI, William. 2011. *Statistics for Engineers and Scientists.* New York: McGraw-Hill.

PAN, Steve, Kaye CHON and Hiyan SONG. 2008. Visualizing Tourism Trends: A Combinition of Atlas.ti and BiPlot. *Journal of Travel Research* 46: 339-348.

PEARSON, Karl. 1904. *On the Theory of Contingency and Its Relation to Association and Normal Correlation.* London: Drapers' Company Research Memoirs Biometric Series.

ROHANI, Ghazaleh. 2010. *Designing Garden and Green Environment.* Tehran: Farhang e Jame (in Persian).

RYAN, Grey W. and H. Russel BERNARD. 2000. Data Management and Analysis Methods. In *The Handbook of Qualitative Research*, N. K. Denzin and Y. S. Lincoln, eds. London: Sage.

SEIDEL, John and Udo KELLE. 1995. Different Function of Coding Data in the Analysis of Textual Data. In *Computer-aided Qualitative Data Analysis: Theory, Methods, and Practice*, U. Kelle, ed. London: Sage.

SHAHCHERAGI, Azadeh. 2010. *Paradigms of Paradise.* Tehran: Jahad Daneshgahi.

SIN, Chi Hoong. 2008. Teamwork Involving Qualitative Data Analysis Software: Striking a Balance Between Research Ideals and Pragmatics. *Social Science Computer Review* 26, 3: 350-358.

STRIJBOS, Jan W., Rob L. MARTENS, Frans J. PRINS and Wim M. G. JOCHEMS. 2006. Content Analysis: What Are They Talking About? *Computers and Education* 46, 29-48.

TAHAMI, Dariush. 1928. *National Garden. Tahami Photo-Grammetry Archive.* Tehran. Iran.

TURNER, Tom. 2005. *Garden History: Philosophy and Design 2000 BC-2000 AD.* London and New York: Spon Press.

WAHBA, Sherine mohy El Dine. 2010. Friendly and Beautiful: Environmental Aesthetics in Twenty-First-Century Architecture. *Nexus Network Journal* 12, 3: 459-469.

WEITZMAN, Eben and Matthew B. MILES. 1995. *Computer Programs for Qualitative Data Analysis: A Software Sourcebook.* London: Sage.

ZAPATA-SEPÚLVEDA, Pamela, Felix LÓPEZ-SÁNCHEZ and María Cruz SÁNCHEZ-GÓMEZ. 2011. Content analysis research method with NVivo-6 software in a PhD thesis: an approch to the long-term psychological effects on chilean ex-prisoners survivors of experiences of torture and imprisonment. *Quality and Quantity* 46, 1: 379-390.

About the authors

Farah Habib is Associate Professor of urban design and research director of the Faculty of Art and Architecture, Islamic Azad University, Science and Research Branch of Tehran. She is director in charge of *International Journal of Environmental Science and Technology* and Editor-in-Chief of *Hoviatshahr Journal.* She has published more than 100 papers. Her research interests include urban morphology, urban history, and architecture.

Iraj Etesam is Professor of architecture and chairman of Department of Art and Architecture at Islamic Azad University, Science and Research Branch of Tehran. He is Professor Emeritus of Architecture and Urban Design at University of Tehran. He was also a professor of Architecture and Urban Design, University of Washington, Seattle, USA. He has published more than 40 papers. His research interests include contemporary architecture, urban design of Iran, and architecture education.

S. Hadi Ghoddusifar is a Ph.D. candidate of architecture at Islamic Azad University, Science and Research Branch of Tehran. He published more than 14 papers in Farsi and English. His research interests include research methods in architecture, landscape architecture, and interdisciplinary research methods.

Nahid Mohajeri is an architect and urban designer and geographer. She got her master's degree in architecture engineering and her Ph.D. in urban design in 2007 from Islamic Azad University, Science and Research Branch of Tehran. She taught several courses in architecture as a lecturer from 2002 to 2008, and from 2008 she has been an Assistant Professor. She went to University College London (UCL), Department of Geography, in 2009 to undertake further studies (on leave from her post). Her research interests include exploring the complex geometric patterns of cities, city shape and urban morphology, as well as using quantitative methods and physical principles in architecture, urban studies, and geography.

Francisco Roldán

C/ Santo Sepulcro 61
18008 Granada SPAIN
froldan@ugr.es

Research

Method of Modulation and Sizing of Historic Architecture

Keywords: Dimensions, geometric systems, ad quadratum, incommensurable ratios, irrational numbers, measuring systems, metrology, number series, patterns, polygons, prime numbers, proportional analysis, regular polygons, rule-based architecture, sequences, series, silver mean, squaring the circle, symmetry

Abstract. Both the base duodecimal arithmetic and geometrical procedures derived from the diagonal of a square were recurrent resources in the design and construction of past architecture. The hypothesis of a double metric scale justifies the modulation size in buildings throughout a long historical period, using a simple and practical procedure whose fundamentals and characteristics are presented here. Validation by other researchers of the method proposed, would be a milestone in the history of the proportion of architecture, a step towards gaining knowledge of a common metric system used since ancient times in the construction of important buildings.

Introduction

What I have proposed is a method for modulation and dimensioning buildings of the past. It is based on a dual scale, a duodecimal arithmetic base and an octagonal graphic base. The modules are determined on two orthogonal axes of reference.

Duodecimal arithmetic base

The classical system of anthropometric units, which is in base 12, has a remote origin, as does the sexagesimal – base 60 – system, used to measure angles and time. Historically it has been used extensively in determining distances, areas and volumes until it was adopted by of the current international metric system.[1] It survives today in the metric imperial system and in isolated local areas.

The base magnitude of length L – and therefore the magnitudes derived from area L^2 and volume L^3 – is defined from a anthropometric base unit value corresponding to the height of man. Derived units are determined by fractions of the duodecimal arithmetic base (prime factors of the base 2, 3).

$$L(2,3) = 1, 1/2, 1/3, 1/4, 1/6, 1/8, 1/9, 1/12,...$$

The different civilizations that have used this system never established their base unit as full anthropometric module (the fathom) – perhaps because it was too big and impractical – but rather as a fraction of it (the foot, the cubit or the yard). The palm is a common sub-multiple (1/24), which is divided into 4 digits, or 3 inches in the uncial system (fig. 1).

The dimensional value of the base unit was fixed for each power or government – arbitrarily, it appears – in the same way that the value of the units of the weight or currency system was established. For this reason, the dimensions of the units of different historical and local areas (fig. 2) do not coincide, although in each area the proportional system is similar.[2]

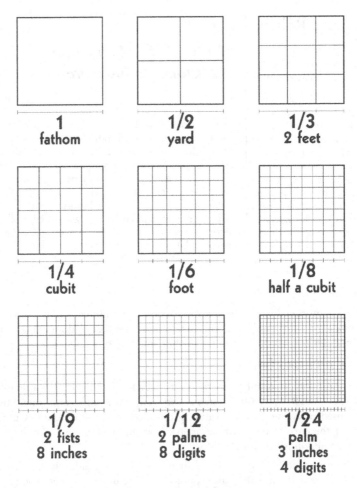

Fig. 1. Duodecimal frames

Unit	Unidad	Fraction	SI	Roma	Imperial	Castilla
fathom	Braza	1	200.00	177.42	182.88	167.181
yard	Vara	1/2	100.00	88.71	91.44	83.59
cubit	Codo	1/4	50.00	44.36	45.72	41.80
foot	Pie	1/6	33.33	29.57	30.48	27.86
	Curata	1/8	25.00	22.18	22.86	20.90
	Sesma	1/12	16.67	14.79	15.24	13.93
	Ochava	1/16	12.50	11.09	11.43	10.45
fist	Puño	1/18	11.11	9.86	10.16	9.29
palm	Palmo	1/24	8.33	7.39	7.62	6.97
inch	Pulgada	1/72	2.78	2.46	2.54	2.32
digit	Dedo	1/96	2.08	1.85	1.91	1.74

Fig. 2. Table of values

Linking units of length with human body parts – as a mnemonic code – facilitates simple arithmetic operations between fractions. Nevertheless, this base system is not operative when divided by other prime factors (1/5, 1/7, 1/11 ...).

Octagonal graphics base

The double square rotated 45° is one of the graphic resources most used throughout the history of architecture. It is known as a measurement tool called Double Egyptian Remen, whose value corresponded to the √2 – square root of two – of a Cubit.

In all medieval and Renaissance treatises[3] the procedure called *ad quadratum* or "in quadrature" is used. This consists in dividing each side of the square in the middle. Connecting these points results in another square that is inscribed and rotated 45°; this new square has an area that is half of that of the first square. If the procedure is repeated with the second square, a third horizontal square is produced, and its sides measure half of those of the first square. Its area is half of that of the second and one fourth of the first, and so on (fig. 3).

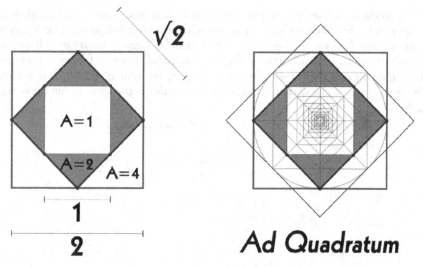

Fig. 3. Ad quadratum

These and other properties, such as those that are used in the DIN-A format, are caused by irrational value √2 – which is obtained from the diagonal of the square –, so the application was often used to duplicate and divide areas as well as to make dimensioned templates to lay out different architectural elements.[4]

$$\mathbf{L\ (\sqrt2) = (1,\ 1/\sqrt2,\ 1/2,\ 1/2\sqrt2,\ 1/4,\ ...)}$$

Modulation procedure

This involves using the duodecimal arithmetic base and octagonal graphic base together. The dimensions of the modules are established by one of the following procedures (fig. 4):

a) Corresponding duodecimal system units;

b) √2 of these values;

c) Both a) and b).

Option c) can be $1+\sqrt2 = \theta$ – the silver mean – or other combinations such as √2–1, $2+\sqrt2$, $1+2\sqrt2$...

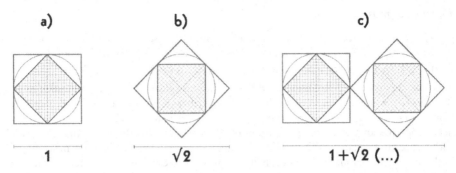

Fig. 4. Procedures

The modulation is static if you use only one of these procedures, and dynamic if you use more than one. Frames originating from orthogonal modulations are static if they use the same procedure on the two axes. These modular frames are dynamic if you use different procedures in each axis. If they are octagonal dynamic frames you can also delimit the two axes rotated 45° by just exchanging the scale of the $1+\sqrt{2}=\theta$ modules. This frame generates the silver rectangles and the Cordovan triangles[5] within the octagon (fig. 5).

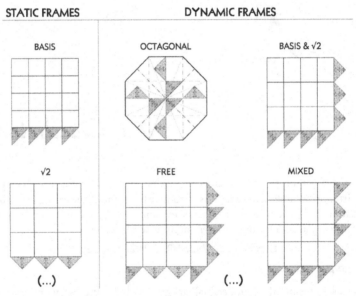

Fig. 5. Typology of frames

Total and partial linear dimensions dynamically modulated are expressed by the positive value of:

$$M = (a \pm b\sqrt{2}) / (2, 3)$$

where a and b are positive integer numbers, and $/ (2, 3)$ is a fraction of the duodecimal base. These dimensions cannot be divided into integer values because they present irrational values.

In statistic modulation, a or b is zero.

Adjustment of limits: Residues

Modulations that approximate overall dimensions determined by another, different modular method are allowed. The differences or irrational remainders accumulate symmetrically at the ends, and are called Residues (fig. 6).

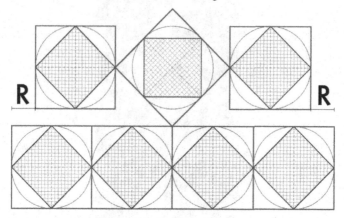

Fig. 6. Residues

Dynamic approaches

Dynamic modulations offer a practical and operational approach – less than 1% error – both fractions of prime factors not present in the duodecimal system (1/5, 1/7, 1/11, etc.) as in other irrational values used in architecture (Phi, Φ: golden number or divine proportion; Pi, π: ratio of length and radius of the circle; Cordovan proportion, c: ratio between the radio and the side of an octagon, or between two unequal sides of a Cordovan triangle).

Fig. 7 shows correspondences of the combination 1+√2 with the approaches to 1/5, Cordovan proportion and Φ, and 2+√2 to 1/7 and π.

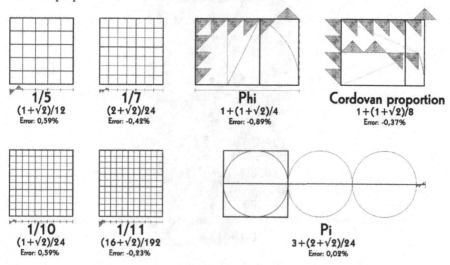

1/5
(1+√2)/12
Error: 0,59%

1/7
(2+√2)/24
Error: -0,42%

Phi
1+(1+√2)/4
Error: -0,89%

Cordovan proportion
1+(1+√2)/8
Error: -0,37%

1/10
(1+√2)/24
Error: 0,59%

1/11
(16+√2)/192
Error: -0,23%

Pi
3+(2+√2)/24
Error: 0,02%

Fig. 7. Approximation

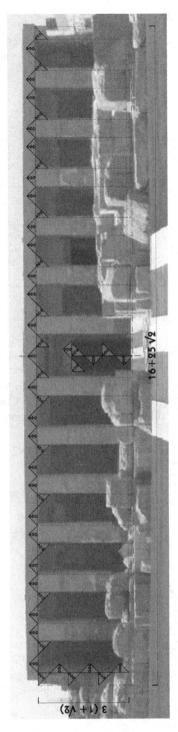

Fig. 8. Temple of Sethy I

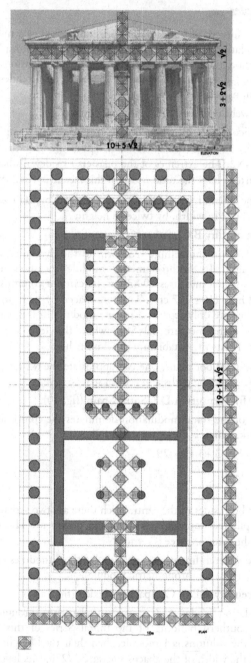

UNIT	fathom	yord	cubit	foot	palm	inch	digit
Unidad	BRAZA	VARA	CODO	PIE	PALMO	PULGADA	DEDO
FRACTION	1	1/2	1/4	1/6	1/24	1/72	1/96
CENTIMETERS	178,97	89,49	44,74	29,83	7,46	2,49	1,86

Fig. 9. The Parthenon

Application Examples

Photogrammetric surveys of buildings are preferably used as a metric reference base, over which the modulations are produced by CAD software (Autocad). Use is also made of ortho-projected photographs, and overall documented dimensions. If dimensions are not known we cannot determine the metric values of the modules; however, we can establish the proportions between them.

Temple of Sethy I (twelfth century B.C), Adibos, Egypt (fig. 8).

The pilaster portico design of this temple corresponds to dynamic modulation. The width of each pillar is $\sqrt{2}$, and is spaced $1+\sqrt{2}$, except at the widest central space between columns, which is $2+\sqrt{2}$. The extremes of the pilasters or sidewalls have a thickness of 1. Because there are thirteen bays the total width of this facade is $16+25\sqrt{2}$.

The height of the pillars from the base is determined by $3(1+\sqrt{2})$. In addition, the main door measures $1+\sqrt{2}$ in width, by twice its height.

The Parthenon (fifth century B.C), Athens, Greece (fig. 9).

Debated proportions in this temple[6] conform to the dynamic modulations both in floor plans and height. Its anthropometric module measures 178.98 cm, which corresponds to a foot which measures 29.83 cm. Its exterior columns are one module in diameter and $3+2\sqrt{2}$ high (1043.17 cm). They are spaced by $\sqrt{2}$ of the module, except at the far ends where they are only separated by one module. In this way the main facades, with eight columns each, measure $10+5\sqrt{2}$, while the seventeen side columns are $19+14\sqrt{2}$. Above them rests the entablature of $\sqrt{2}$ units high.

Through the restored floor plans, one can appreciate the $\sqrt{2}$ rate arrangement of the interior columns that shape the paths and central spaces.

Arch of Medinaceli (first century A.D.), Soria, Spain (fig. 10).

The regular grid of the static modulation dominates the composition of this Roman civil engineering work, which is dimensioned as follows:

Width: 3 +3 +3 +11 +3 +3 +3 = 29

Height: 8 +9 +3 = 20

Bottom: 4

Furthermore, on both sides of the central arch there are two separate elements formed of two Corinthian pilasters topped by a triangular pediment. The total width of each element is $3\sqrt{2}$ and the interior space is set to $2\sqrt{2}$.

According to the general dimensions published,[7] these modules correspond to cubits value 45.29 cm (181.16 cm anthropometric module).

Xcalumkin (eighth century A.D.), Campache, Mexico (fig. 11).

Dynamic modulation is detected in various elements through the analysis of a photograph of the southern building of the Initial Series of this Mayan city. If the diameter of each of the columns is 1 module, then their total height equals 3, their axes are spaced in $2+\sqrt{2}$, the width of the abacus measures $\sqrt{2}$ and its height is one quarter of $\sqrt{2}$.

Fig. 10. Arch of Medinaceli

Fig. 11. Xcalumkin

Madinat Al-Zahra (tenth century A.D.), Cordoba, Spain (fig. 12).

The facade of the Salón Rico[8] of Caliph Abd al-Rahman III in this palatial city is composed of an arcade of five arches separated by the √2 anthropometric module, corresponding to a 42.48 cm cubit. It is framed by a decorative module, the total width being 2 +5√2 and the height 4.

On the floor the interior spaces are sized in straight modules, except through the arcade of the Salón Rico, which matches the scale √2. The thickness of the longitudinal walls is dimensioned dynamically (1+√2). The two transverse walls have a width of 2√2 and the measurement of the bottom is 2+√2.

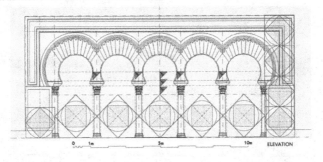

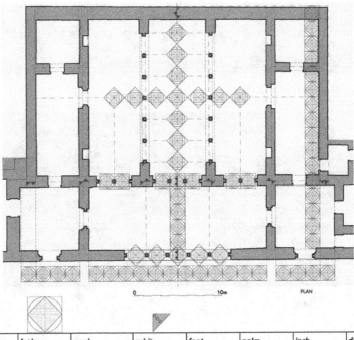

UNIT	fathom	yard	cubit	foot	palm	inch	digit
Unidad	BRAZA	VARA	CODO	PIE	PALMO	PULGADA	DEDO
FRACTION	1	1/2	1/4	1/6	1/24	1/72	1/96
CENTIMETERS	169,92	84,96	42,48	28,32	7,08	2,36	1,77

Fig. 12. Salón Rico in Madinat Al-Zahra

Cuarto Real De Santo Domingo (thirteenth century A.D.), Granada, Spain (fig. 13).

This is an emblematic Nasrid palace earlier than and very similar to those of the Alhambra. The results of the study of their modulation[9] – of which only the general dimensions are presented here – have enabled the author to establish the characteristics of the metrology of classical architecture described in the present work.

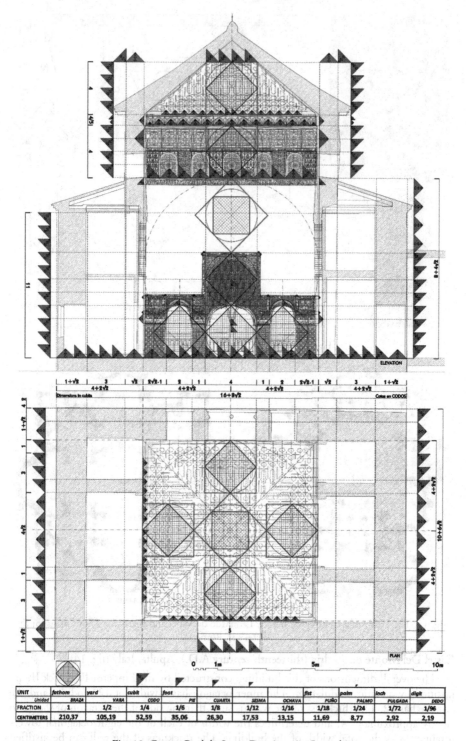

Fig. 13. Cuarto Real de Santo Domingo in Granada

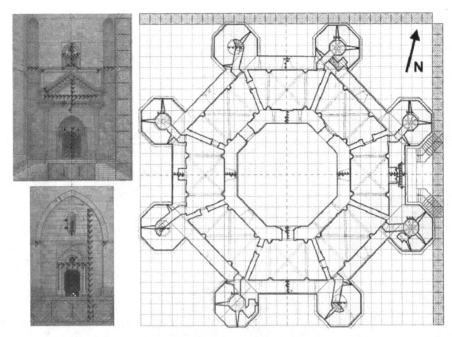

Fig. 14. Castel del Monte on Andria, Apulia

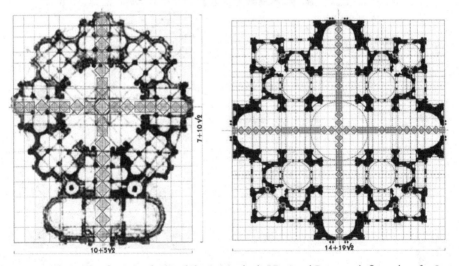

Fig. 15. The Study of a Central Church by Leonardo da Vinci and Bramante's floor plans for St. Peter's in Rome

Castel Del Monte on Andria (thirteenth century A.D.), Apulia, Italy (fig. 14)

The overall dimensions of this building, constructed by the Emperor Frederick II, are justified by a static modulation. The inner octagon has a width of 12 modules and the outside has 22. The thickness of the walls is such that the side of the octagon of the patio closely approximates 4, and determines the width of each of the eight octagonal towers totaling 28 as the total width of the building. The thickness of the wall can be justified

statically by 29/24, or dynamically by $(1+\sqrt{2})/2$. Other dynamic modulations are detected in the widths of the holes of the main facade, the skylight over the balcony of the courtyard, and other elements. The framed part of this balcony is an exact golden rectangle.

The study of a centrally planned church by Leonardo da Vinci (ca. 1488 A.D.) and Bramante's floor plans for St. Peter's in Rome (1505-1506 A.D.) (fig. 15)

Both central compositions are substantiated by dynamic modular frames. The scale $\sqrt{2}$ can be appreciated in determining the axes and routes of each sacred space.

Conclusions

The deduced metric system coincides with the units derived from anthropometric duodecimal canon, but includes one other scale in proportion to $\sqrt{2}$. It can be used separately or together. Since the number 2 is contained in the $\sqrt{2}$ series, we can express it like this:

$$L(\sqrt{2},3) = (1, 1/\sqrt{2}, 1/2, ...) (1, 1/3, 1/9, ...)$$

In this way, it is compatible with the static scale in regular grid, and in addition, with all the dynamic *ad quadratum* designs generated from octagonal symmetry, which are traced with total metric control on the four axes of symmetry. One example of this is the elaborate works of Hispano-Muslim decoration proper to the "Mudéjar style".

Using a modulation method and a very simple and practical layout, this method also provides another possible variety of combinations, good approximations to other architectural constants, and facilitates prefabrication in the workshop.

This explains the principle of proportionality between the parts, in works from antiquity.

There is more use of the $\sqrt{2}$ scale in palaces of royalty, and especially in temples and sacred motives.

The double scale system described here represents an effective tool for documenting and analyzing the architectural heritage, and its application seems advisable for in all historical studies.

Notes

1. Throughout history, people have used other mathematical, numerical systems: the binary numeral system, which is in base 2; the ternary system, which is in base 3; the quinary system, which is in base 5; the octal system, which is in base 8; the decimal system, the most common system in use today, which is in base 10; the hexadecimal system, which is in base 16; the vigesimal system, which is in base 20. However, in arithmetical terms, the duodecimal system and the sexagesimal system admits a greater number of fractions, while in geometrical terms, they are compatible with square and triangular-hexagonal frames.
2. The observed variation in the size of the different units was a great disadvantage of the classical system of proportions, but it was not the only one. For example, Filarete [1990: 52-53], in his treatise on architecture, describes a system of different modules defined by Vitruvius [2007]. He highlights the language problem that occurred when using similar names for units, which corresponded to different fractions of the anthropometric module. To this, we must add that even in the same area, one unit would have different size depending on the type of merchandise to be measured. The variety of measures and units has always favoured a host of local names and misunderstandings in the original texts, translations and interpretations. According to Vitruvius the parts of a temple should be in proportion to each other, as they are in a well-formed man. The modules could be many sizes and denominations, however in the

construction of a building there should be metric unity in order to carry this out. Each work can have its own and unique module, as each man has a different height.

3. Among the most ancient treatises we can cite those of Matthäus Roritzer, Hans Schmuttermayer and Lorenz Lechler, by using the procedure thoroughly. The Renaissance treatises prefer to use measurable fractions, but continue to use it for certain paths [Ruiz de la Rosa 1987].

4. The ratio between different parts of buildings according to the √2 has been studied by Violet-le-Duc, Jay Hambidge, Tons Brunes [1967], Carol M. Watts [1987], Rafael Vila [1997] and Jay Kappraff [2002], among others.

5. The Cordovan triangles are present in the octagonal dynamic frames, as Tomás Gil López [2012] verifies. The Cordovan rectangle or Cordovan proportion is proposed by Rafael de la Hoz [1973] as ratio of several rectangular architectural bodies of the Great Mosque and other buildings in Córdoba, Spain.

6. Luis Moya, in his *Relación de diversas hipótesis sobre las proporciones del Partenón* [1981], examines the proposals of Barnacles Nicolas, Jay Hambidge, Viollet-le-Duc, Georges Tubeuf, Lesueur, Charles Chipiez, C. J. Moe, Maruis Clayet-Michaud, Henri Trezzini, D. R. Hay, August Thiersch, Alexander Speltz, Zeysing, Mossel, Matila C. Ghyka, Ernst Neufert, Wedelphol, Hans Plessner, Funck-Heller, Otto Hertwing, Odilo Wolff, Karl F. Wieninger, Victor D'Ors, Uhde and other theorists. Moya begins his conclusions as follows: "A close look at the diverse hypothesis with respect to the proportions of the Parthenon has demonstrated that no system can explain, simultaneously, the two aspects of the problem: First of all, the real dimensions of the measurements at present, and secondly, how they were obtained in the construction process" (my translation). Other studies have appeared, usually emphasizing Matila C. Ghyka's theory of the Divine Proportion, spearheaded by Le Corbusier, as a counterpoint to highlight those proposed by Tons Brunes [1967] and Anne Bulckens [2002].

7. Overall dimensions published by Lorenzo Abad Casal [2002].

8. Photogrammetry performed by Antonio Almagro [2011].

9. Francisco Roldán [2011] with photogrammetry performed by Antonio Almagro and Antonio Orihuela [1997].

References

ABAD CASAL, L. 2002. *El Arco de Medinaceli (Soria, Hispania Citerior)*. Madrid: Real Academia de la Historia.

ALMAGRO, A. 2011. *Planimetría de Madinat Al-Zahra*. Consejo Superior de Investigaciones Científicas (CSIC). Granada: Real Academia de Bellas Artes de San Fernando.

ALMAGRO, A. and ORIHUELA UZAL, A. 1997. *Propuesta de intervención en el Cuarto Real de Santo Domingo (Granada)*. Valencia: Loggia.

BRUNES, T. 1967. *The Secrets of Ancient Geometry – and its use*. Copenhagen: International Science Publishers.

BULCKENS, A. 2002. The Parthenon Height Measurements: The Parthenon Scale with Roots of 2. *Symmetry: Art and Science* 2, 1-4: 219-229.

FILARETE. 1990. *Tratado de Arquitectura*. P. Pedraza, ed. Vitoria: Ephialte.

GIL LÓPEZ, T. 2012. The Vault of the Chapel of the Presentation in Burgos Cathedral: "Divine Canon? No, Cordovan Proportion". *Nexus Network Journal* 14, 1: 177-189.

HOZ, R. 1973. La proporción Cordobesa. *Actas de la quinta asamblea de instituciones de Cultura de las Diputaciones*. Ed. Diputación de Córdoba.

KAPPRAFF, J. 2002. *Beyond Measure: Essays in Nature, Myth, and Number*. Singapore: World Scientific.

MOYA BLANCO, L. 1981. Relación de diversas hipótesis sobre las proporciones del Partenón. Madrid: *Boletín de la Real Academia de Bellas Artes de San Fernando* 52: 25-156.

PACIOLI, L. 1987. *La Divina Proporción*. A. M. González Rodríguez and J. Calatrava, eds. Torrejón de Ardoz, Madrid: Akal.

ROLDÁN, F. 2011. *La Escuadra Sagrada*. Madrid: Bubok Publishing S.L.

RUIZ DE LA ROSA, J. A. 1987. *Traza y simetría de la Arquitectura: En la Antigüedad y el Medievo*. Sevilla: Universidad de Sevilla.

VILA RODRÍGUEZ, R. 1997. Estudios compositivos de algunas basílicas paleocristianas de la España romana de los siglos IV - VI. *Antigüedad y Cristianismo: Monografías Históricas sobre la Antigüedad Tardía* **14**: 489-500.

VITRUVIUS. 2007. *Los diez libros de Arquitectura.* D. Rodríguez Ruiz and J. Ortiz y Sanz, eds. Torrejón de Ardoz, Madrid: Akal.

WATTS, C. M. 1987. A Pattern Language for Houses at Pompeii Herculaneum and Ostia. Ph.D. dissertation, University of Texas at Austin.

About the author

Francisco Roldán is Technical Architect (1986), Architect (1994), Professor of CAD in the Higher School of Architecture of the University of Seville (1990–1994). He holds a Diploma of Advanced Studies at the University of Granada (2000). He is a Professor of Descriptive Geometry of the Higher School of Architecture of the University of Granada (2008-2009) and Professor of Construction and Thesis Work in the Higher School of Engineering Building of University of Granada (2009–2011).

Riccardo Migliari

'Sapienza' Università di Roma
Dipartimento di Storia, Disegno
e Restauro dell'Architettura
Piazza Borghese, 9
00186 Rome, ITALY
riccardo.migliari@uniroma1.it

Keywords: Descriptive geometry,

Research

Descriptive Geometry: From its Past to its Future

Abstract. Descriptive geometry is the science that Gaspard Monge systematized in 1794 and that was widely developed in Europe, up until the first decades of the twentieth century. The main purpose of this science is the certain and accurate representation of three-dimensional shapes on the two-dimensional support of the drawing, while its chief application is the study of geometric shapes and their characteristics, in graphic and visual form. We can therefore understand how descriptive geometry has been, on the one hand, the object of theoretical studies, and, on the other, an essential tool for designers, engineers and architects. Nevertheless, at the end of the last century, the availability of electronic machines capable of representing three-dimensional shapes has produced an epochal change, because designers have adopted the new digital techniques almost exclusively. The purpose of this paper is to show how it is possible to give new life to the ancient science of representation and, at the same time, to endow CAD with the dignity of the history that precedes it.

Introduction

Descriptive geometry is the science that Gaspard Monge systematized in 1794 and that was widely developed in Europe up until the first decades of the twentieth century. The main purpose of this science is the representation, certain and accurate, of shapes of three dimensions on the two-dimensional support of the drawing; while its chief application is the study of the geometric shapes and their characteristics, in a graphic and visual form. We can therefore understand how descriptive geometry has been, on the one hand, the object of theoretical studies, and, on the other, an essential tool for designers, engineers and architects.

Nevertheless, at the end of the last century, the availability of electronic machines, capable of representing three-dimensional shapes, has produced an epochal change, because the designers have adopted the new digital techniques almost exclusively. Furthermore, mathematicians seem to have lost all interest in descriptive geometry, while its teaching in universities has almost disappeared, replaced by a training in the use of CAD software, which mainly has a technical character.

The purpose of this paper is to show how it is possible to give new life to the ancient science of representation and, at the same time, to endow CAD with the dignity of the history that precedes it.

This result may be achieved by verifying and validating some fundamental ideas:

- the idea that descriptive geometry is set within a historical process much wider than the Enlightenment period, a process which goes from Vitruvius to the present day, and that it therefore includes both the compass as well as modern digital technologies;

DOI 10.1007/s00004-012-0127-3; *published online* 21 September 2012

- the idea that, to the graphic representation methods (perspective, the method of Monge, axonometry, topographic mapping) can today be added the digital methods that are implemented in computer applications (mathematical representation, numerical or polygonal representation);
- the idea that the synergy between the calculation and the visualization of the shape offered by the digital systems, may provide simpler and more general solutions to the classic problems of descriptive geometry;
- the idea that from this view of their historical foundations digital applications should receive a stimulus towards the unification of the terms used for the procedures, the shapes and the operations that the applications offer to the user.

Research groups of five different Italian universities are working on this topic (Rome, Milan, Genoa, Venice and Udine). Here, we present the main lines of this research and the first results.

Apparently, in architectural practices, CAD has replaced descriptive geometry as a tool for the representation of three-dimensional shapes. In universities, the teaching of descriptive geometry is disappearing. The mathematicians have not cared about these studies since the first decade of the last century. Does a future exist for descriptive geometry? Is it possible to give this ancient science a new life?

A look back at history

If we would like to glimpse what the future of a science may be, we must recall its past, since in its past was traced the path that leads, today, towards the future.

In 1794 Gaspard Monge explained, in a course of lectures at the École Normale in Paris, the fundamentals and the first applications of a discipline that then seemed to be totally new; he gave it the name it is known by today: *Géométrie Descriptive*. In the fevered atmosphere of the French Revolution, only very few intellectuals dared to re-evaluate the originality of Monge's work. Joseph Louis Lagrange did it, with plenty of irony, after having attended one of these lessons, exclaiming: *Je ne savais pas que je savais la géométrie descriptive!* ('I didn't know that I knew descriptive geometry'). [1] But others understood Lagrange's words as a proof of the clarity of Monge's exposition and pretended not to understand. Michel Chasles also tried to place *Géométrie Descriptive* in its historical perspective, [2] and he did it with reasoned arguments, but his efforts were not enough to prevent the image of Monge, 'creator' of the science that he baptized, from reaching us in the present time.

Actually, as everyone who has studied history of art knows, descriptive geometry has much older roots. Therefore I think that we should write *Géométrie Descriptive*, in French, when we refer to the science developed by Monge and his school, and write simply 'descriptive geometry' when we allude to the geometric science of representation in its centuries-old journey.

Descriptive geometry teaches to construct and represent shapes of three dimensions and, with these, the objects of all kinds of artistic, planning or production activities. These representations are drawings that are constructed following a geometric code, which permits us to move from the two-dimensional space of the representation to the three-dimensional space of the physical object. Thanks to its ability to create bi-univocal relations between the real space and the imaginary space of the drawing, descriptive geometry also lends itself to many applications, which range from the study of the properties of surfaces, to the creation of spaces and illusory visions.

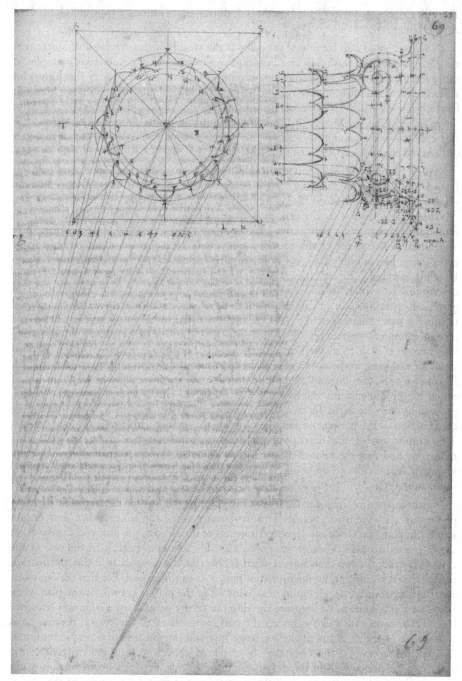

Fig. 1. Fol. 69r of Piero della Francesca's manuscript *De prospectiva pingendi* preserved in Bordeaux. The page represents, in orthogonal projections, the Italic capital, the point of view, the picture plane and a few projection operations of significant points of the capital

Thus, descriptive geometry has strengthened, during the centuries, a fruitful link between art and science.

Now, we could talk for a long time about the wealth of this synthesis of abstraction and manual skill, of reasoning and intuition. And we could cite numerous examples of the outcomes of this synergy throughout the history of art, as well as the history of science. But this discussion, even if fascinating, would turn our attention away from out first objective, which is that of demonstrating that descriptive geometry was not 'invented' by Gaspard Monge in 1794 and that, instead, it has a much older history. For this end, it will be enough to illustrate an example, which will also enable us to better understand the *modus operandi* of the discipline.

Three centuries before Monge (around 1480) Piero Della Francesca composed his famous *De Prospectiva Pingendi,*[3] a treatise that teaches how to construct perspectives of objects of three dimensions, using representations of the same objects in plan and elevation. The treatise is divided into three books and contemplates two different methods of constructing perspectives. In the second and third book, in particular, there are minute descriptions of the operations necessary to construct object of remarkable complexity, like buildings, a cross vault, an attic base, a torus (the *mazzocchio* of Paolo Uccello), an Italic capital (fig. 1), an apsidal half-dome divided in coffers. The description is written in an algorithmic form, or better, as a well organized list of graphical operations, all practicable, which, based on certain data, lead to the desired result: the representation of the object as it is perceived by the eye of a man placed in a certain point of observation. If we compare the amount of objective and operating information contained in the text with the number of signs that appear in the supplied small illustrations, we become aware that the graphical description is much less detailed. In other words, the illustration supplied to the text is a mere allusion to a drawing of much bigger dimensions; this is what Piero observes and reconstructs for the reader, proceeding step by step. In fact, Piero enunciates a theory, which is his method of construction of the perspective, and he supports this theory with a series of experiments. The minuteness of the description of each experiment serves to ensure that it is repeatable and that the related theory is thus validated. To be persuaded of what I am saying, it is sufficient to draw one of these drawings again, for instance, the one of the Italic capital (fig. 2).

Without entering into details, I will only examine the flow of Piero's work. In a first phase our scientist-artist explains how to construct the capital in *width and height,* namely in plan and elevation. In other drawings of the treatise, the plan and the elevation are connected to each other by what we today call 'reference lines'.[4] In the case of the capital, instead, the two drawings are separate, because the complexity of the construction is such that it requires using the format of the paper to the utmost. For that reason, when we have to construct the elevation of a point when the plan is known, or the plan when the elevation is known, we measure the distance of the point from a common reference line. This proves that Piero connects the two 'projections' of the object, because he is fully aware of the meaning of the reference line and not because of a simple intuition. Piero, in other words, conceives the object placed above the plan and in front of the elevation, as Monge will do three centuries later. In this meticulous construction work, the genesis of the representation of the geometric entities and the genesis of the object, proceed hand in hand. This is the *modus operandi* typical of descriptive geometry: the image emerges as the object takes shape in the mental space of the designer and only if the designer is able to give the object a shape. The geometric construction and the simulation of the physical construction are simultaneous.

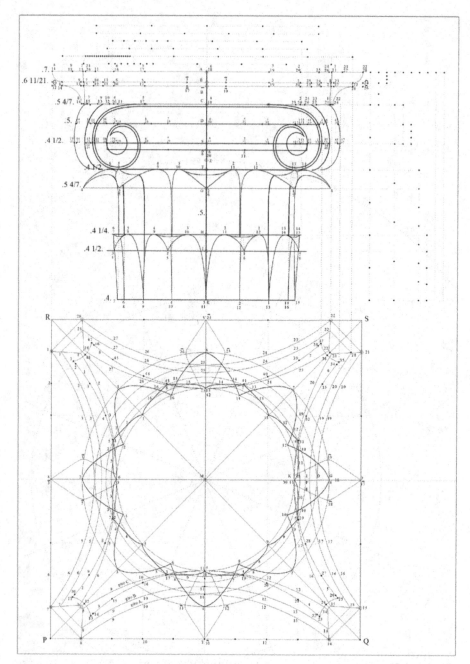

Fig. 2. This drawing, which is readable only in in-folio format, represents the graphic transcription of operations related to the construction of the capital, meticulously described by Piero from fol. 49v to fol. 52r of the manuscript preserved in the Biblioteca Palatina in Parma

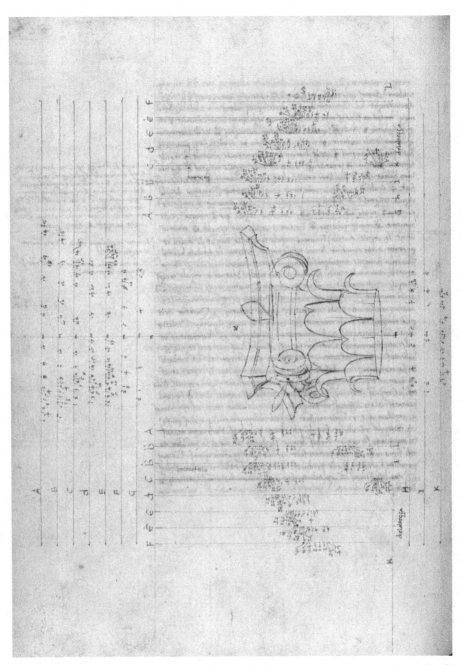

Fig. 3. Fol. 72v of the manuscript preserved in Bordeaux, in which Piero represents the result of the projection operations previously carried out on the capital and experimentally verifies the perspective vision

In the second phase, after having constructed the capital, Piero determines in the space the point of view, the plane surface of the picture plane that will contain the perspective, and he carries out the operations of projection and section that give rise to the perspective illusion. This is another extraordinary moment in the history of descriptive geometry, because, perhaps for the first time, here are described and realized geometric operations carried out in a three-dimensional virtual space, which arises from a representation in plan and elevation. In other words, just as Monge will do three centuries later, Piero uses the two associated orthogonal projections not only in order to generate two significant images, but also to work on the model that these images are able to evoke.

The operations described will generate a crop of experimental data, which are the 'coordinates' of the points in which the visual beams meet the picture plane.

In the third and final phase, the data collected in the second phase are methodically written on paper by means of special strips of paper and wood and produce, we would say today, a point cloud that describes the perspective of the capital (fig. 3).

Topicality of descriptive geometry

At this point, I think it is necessary to define a question that is of an ethical nature, so as not to lead to misunderstandings: I don't want to belittle Gaspard Monge's role in the history of descriptive geometry, nor, even less, in the history of science. I don't want to subtract from Monge any of the credit that was given to him. I only wish to give back to descriptive geometry its past as well as its functionalities, and to show how these functionalities are still topical today.

In fact, if we consider *Géométrie Descriptive* as an offspring of the Age of Enlightenment and of the Industrial Revolution, we might also legitimately claim that, in comparison with the era of the computer, this science has outlived its time. But, if we instead consider descriptive geometry for what it is in itself – a science that is rooted in the past, even before Piero's time, and rooted in the art of thinking and creating space, more than connected to the techniques of production –, then we will realize that this science has not yet exhausted its life cycle and that it still deserves to be considered, studied and developed.

Today, as everyone knows, computers enable us to create three-dimensional models of objects and of geometric shapes. They can also automatically generate the Mongian projections, and not only, of those objects. This technique is called 'computer-aided design' (CAD).

Nowadays, as in 1794, crowds of students attend our universities to learn the art of imagining the objects of the future: houses, furnishings, cities, machines. Now, as then, we are faced with the problem of providing them with theoretical and operating tools useful to practise this art of the invention and pre-figuration of space. Can the CAD take the place of descriptive geometry, or is it instead descriptive geometry that has to integrate the CAD among its tools? And, if we would like to carry out this integration, how could we realize it?

The tools of descriptive geometry

I think that the answer to the first question is clear. Computers are tools. They are sophisticated tools, but analogous to the straightedge and compass, which were, for years, the only mechanical tools admitted in the study of geometry. At this point two questions arise: the first concerns, in general, the role of the tools in geometry; the second whether

it is right to go beyond, once and for all, the constraint that was set by classical geometry on the exclusive use of straightedge and compass.

The role of the tools is tied to the experimental character which is present in geometry in general, and particularly in descriptive geometry. These sciences are founded on the vision and the graphical verification of the shapes while created and studied. When, within the context of an abstract reasoning, we introduce the idea of a right angle or of any shape whatsoever, such as a cone or a round hyperboloid, these ideas immediately give rise to the images that are connected to them. It is therefore impossible to reason about geometry without, at the same time, seeing with the mind what we are reasoning about. And what we imagine can also be drawn and seen clearly and shown to others. Naturally, the outcome of a graphical experience cannot, in itself, be a guarantee of scientific truth, which can only be obtained by means of a correct logic, but the graphical experience support the reasoning and, above all, stimulates the reason with its allusions.

Monge himself, when defining the second aim of *Géométrie Descriptive*, which is that of studying the properties of the shapes, says that the geometric experience offers numerous examples of the passage 'from the known to the unknown' (*du connu à l'inconnu*).[5] This affirmation, if we stop a moment to consider it, is surprising. Surprising, because we would expect that the graphic representation of a geometric idea is a way to change this from an embrionic condition of an intuition into the certainty of the image, namely, something that we can see and nearly touch. We would expect, therefore, a passage from the unknown to the known. Monge, instead, goes beyond this passage and highlights the heuristic character of the graphic experience, that is, the moment in which the genesis of the image, which forms itself right before our eyes, suggests, without making explicit, relations, properties and characteristics that the intuition did not suspect.

If it is therefore right to use the drawing tools in geometry, not only to show and to verify, but also to experiment, it is unavoidable to ask ourselves which tools should be allowable.

Now, as we already said, for centuries these tools were confined to the straightedge and compass. How can we explain this dogma of ancient and modern science? According to me, there is only one possible explanation: straightedges and compasses, for centuries, were the only tools able to guarantee an acceptable graphic accuracy, therefore, an acceptable experimental verification. If not, what other reasons could François Viète have had for rejecting, almost with contempt, the solution given by Adriaan Van Roomen to the Apollonian problem?[6] And yet the solution given by Van Roomen (fig. 4) was simple and general, able to tackle with the same logic the series of complex cases of the problem, and lent itself to being extended to space.

But it had a fault: it made use of geometric loci, the conics, which could not at that time be drawn accurately. These same reasons suggested to Lorenzo Mascheroni a geometry completely solved using only the compasses, because, as he himself says in the preface of his work, *è quasi impossibile ch'essa [la riga] sia così dritta che ne garantisca per tutto il suo tratto della posizione a luogo de' punti, che in essa sono* (It is almost impossible that the rule is so similar to a straight line that it can guarantee that the points which lie on its edge all are aligned).[7] We could go on at length, with these examples of 'experimental' geometry, and up until the present day: you only have to recall the works and the investigating techniques of H. S. M. Coxeter.[8]

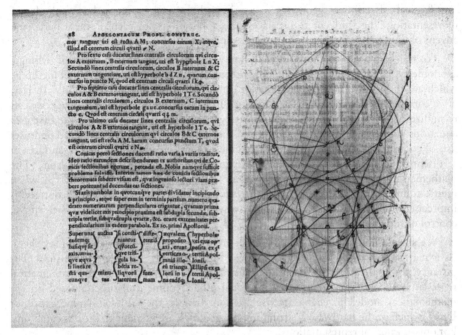

Fig. 4. Adriaan Van Roomen, *Problema Apolloniacum* (1596). In the figure on the right can be seen the hyperbolas that describe the geometric loci of the points equidistant from the given circles and that enable a simple solution of the problem

Well then, as we said, computers are tools. They are tools that, thanks to the synergy with the computation, are able to draw a straight line, a circle, the conics and even much more complex lines, all with the same accuracy. In applications commonly used in industry, this measurement accuracy is on the order of a micron. I would like to recall that traditional technical drawing can achieve, theoretically, the accuracy of two-tenths of a millimetre; thus computers have improved the accuracy of the experiments, which can be performed in geometry, by two size orders.

But, there is something more, because the analogical drawing can only draw lines on plane supports, whereas the digital drawing can draw lines and surfaces in space. Therefore, if (formulating a hypothesis out of its historical context) Piero had had a computer, he could have simplified the second part of his procedure a lot, drawing the projecting lines of the visual pyramid, each one with a single stroke in space. The first and last parts of Piero's experiment, instead, would have kept their laborious character, the first, because it deals with the construction of the capital, which is a problem of curve shapes and skew surfaces, connected by a delicate system of relations, the last, because it translates a discrete system – the point cloud – into a continuous system, with an evident contribution of the interpretation. In all these phases, the role of descriptive geometry is dominant, in spite of the aid of CAD systems, which are purely instrumental.

An outline of a new structure for descriptive geometry: the methods

If, as I believe, descriptive geometry is still the science of representation of space, and computers only a tool at its disposal, we should begin to wonder how the structure of the discipline can and should integrate the new techniques of experimental verification.

Classical descriptive geometry comprises three main parts: the methods of representation, the study of the surfaces, the applications. In each of these parts, the advent of digital drawing has an important role.

To the graphical methods already known, which are the method of Monge, axonometry, perspective and topographic projection, we have to add the digital methods, which are the 'mathematical representation' and the 'numerical representation'.[9] In fact, if we consider the software dedicated to modelling, or better, to drawing in space, we can recognize two main procedures in representing three-dimensional shapes:[10] the first uses equations and thus describes the curves and the surfaces with continuity; the second uses lists of coordinates of points and rules to connect them, and thus describes the surfaces in a discontinuous or discrete way, approximating them with a polyhedron. The mathematical representation is very accurate and is, for this reason, preferred when metric control of the shape is required. The numerical representation, on the contrary, is imprecise, but easy and quick, and this is why it is preferred when a direct, perceptive, control of the shape is required.

We should not confuse these two methods with the applications that employ them. As a matter of fact, all the applications use both methods, in different measures. For instance, the applications dedicated to industrial production mainly use the mathematical representation, but they generate a polygonal model (numerical) superimposed on the mathematical model, to enable its visualization. In fact, the GPUs (Graphic Processor Unit), which in the hardware are handling the graphics, are not able to process equations, but only numerical representations.

In their turn, the applications dedicated to rendering perspectives, the shadows and chiaroscuro, and to generating animations, mostly use numerical representation, but they also have some mathematical functionalities that make it possible to construct the shapes more rapidly, the basis for subsequent modelling operations.

It is easy to define an analogy between graphical representation methods and those of digital representation if we look not so much at the images that they produce as at the use that architects and artists generally make of them. In the case it is necessary to perform a verification of measurements on the shape, as for instance on the dimensions of an environment system, the architect works using plans and elevations; when instead he wishes to study the outcome in the synthesis of an overall perception, the architect uses perspective for the view from the inside and axonometry for the view from the outside of the planned volumes. We can therefore say that the mathematical representation is analogous to the associated orthogonal projections, because it enables the accurate control of the dimensions; while the numerical representation is analogous to perspective, because it enables accurate control of the view of the object.

Just as we teach and prove geometrically the rules necessary to represent on a plane a three-dimensional object, in a way that it can be re-constructed in space, so, in the descriptive geometry of the future, we can teach the rules necessary to represent in space an object using the descriptions, mathematical or simply numerical, that a machine is able to translate into images in real-time.

An outline of a new structure for descriptive geometry: the study of surfaces

In the past, classical descriptive geometry, when working on a plane, made use exclusively of straightedge and compass. For instance, the solution in space of the Apollonian problem was discussed, in 1812, by Louis Gaultier de Tours, in a *Mémoire* of more than hundred pages,[11] in which the theory of the geometric radicals was enunciated

for the first time. All this was intended to ensure that the restrictive use of the circle as a geometric locus is respected. But today, if we accept the use of the conics as well and, in space, the surfaces of revolution generated from them – namely, the hyperboloid, the paraboloid and the ellipsoid – then the Apollonian problem finds a general solution that can be outlined in a few pages.[12] This solution, moreover, has the merit that it can be really carried out, as far as to construct in space the spheres that touch four geometric entities *ad libitum* chosen among points, planes and spheres, as the statement of the problem calls for (fig. 5).

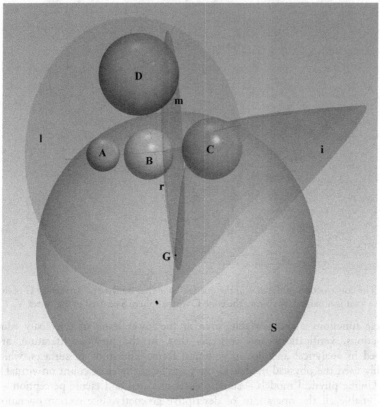

Fig. 5 – The solution of the Problem of Apollonius that uses the mathematical representation applying the method conceived, in the plane, by Adriaan Van Roomen. The sphere **S**, of radius **r**, touches the spheres **A**, **B** and **C** with the inner side of the surface, the sphere **E** with the outer side. The centre **G** is the common point to the hyperbolas intersection of the hyperboloids of revolution **i, l, m**. Sixteen solutions to the Problem are possible, but not all are always feasible

Naturally, the Apollonian problem is only one of the numerous examples that we might give of a new way of studying descriptive geometry, a way that uses digital compasses able to draw second-degree curves and surfaces in space.

But the use of the computer also offers other possibilities, which derive from the synergy between the graphic synthesis and the calculation. For instance, the possibility of calculating the centre of mass of a solid can be applied successfully to the construction of the axes of the quadric cone and to the shapes that have a cone-director, like the elliptic hyperboloid of two sheets.[13] In fact, if we cut the cone with a sphere that has its centre in

the vertex, and we remove the part of the cone that is on the outside of the sphere, the straight line that passes through the vertex and through the barycentre of the shape is the first of the axes, whereas the other two are parallel to the axes of whichever of the ellipses that are obtained cutting the cone with a plane that is perpendicular to the first axis (fig. 6).

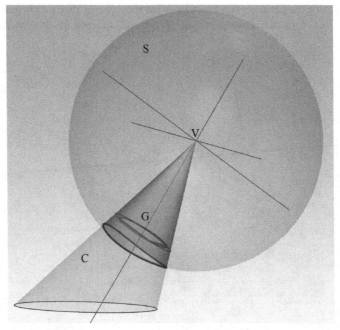

Fig. 6. The construction of the axes of the quadric cone that utilizes the barycentre **G** of the solid that is obtained by cutting the cone **C** with a sphere **S** centred in the vertex **V**

These functions make available, even at the lower levels of university education, constructions, verifications and concepts that, in the previous literature, are only developed in analytical and not in graphical form. The study of surfaces, which very profitably used the physical models in the past,[14] can therefore count on virtual models today. Unlike physical models – static objects of visual and tactile perception –, virtual models enable all the operations of descriptive geometry, like section operations and geometric and projective transformations.

An outline of a new structure for descriptive geometry: the applications

Classical descriptive geometry has a wide range of applications, many of which are tied to the production of objects, others to the production of images, still others to the study of the history of art. In all these cases, the use of digital representation techniques has given interesting outcomes, for both industry and research. Here I only wish to give a few examples, from among the many that can be recalled.

Today, the study of the Gaussian curvature of the surfaces has accurate descriptions in false colours, which are applied to the control of the continuity of surfaces, within every field of industrial production. The construction of developable surfaces enables the realization of the new plastic forms of the buildings designed by Frank Gehry. Perspective is experiencing second youth in its dynamic and interactive formulation,[15] in

which it is no longer the artist who chooses the point of view: it is the observer who changes it, continually exploring the illusory space.

Finally, we cannot forget the contributions that the new digital descriptive geometry has given, and continue to give, to the study of history, as, for instance, in the case of the Roman paintings of the first century, which are evidence of the knowledge of perspective of the ancients.[16]

Descriptive geometry and the education of designers

I do not think I have yet answered thoroughly the question posed at the beginning, namely: Is there still a future for descriptive geometry? Is it possible to give new life to this ancient science?

As a matter of fact, after a superficial review of the question, CAD would seem to be self-sufficient and therefore able to meet the needs of the science, of design and of production, even in the absence of a historical memory. After all, any student who starts to study a modelling application is able, after only a few days, to create three-dimensional shapes.

Indeed, he is able to create them, but not to control them. And it is not by chance that in this empirical approach numerical representation is preferred, with all its approximations. Like clay in a sculptor's hands, the shape represented numerically can be moulded without difficulty, but also without proportions, without measures, without generative laws, in a word, without geometry.

The genesis of a three-dimensional shape, above all when we deal with architecture, is very different. It requires a process that is orderly and guided by reason: the construction.

Let's imagine, for instance, a simple polyhedron like the dodecahedron: it can be constructed using the knowledge of the *mirabili effetti* described by Luca Pacioli,[17] or also moving from the plane to space the development of six of its twelve faces, and then generating the others by symmetry. In both cases, the construction requires a knowledge of inner relations, of rotation operations, of projective relations, all of which belong to descriptive geometry and to its history, and which in no way can be substituted by CAD, because CAD is not a science, but a technique.

We can also mention Piero della Francesca's complex experiment. The construction of the capital is divided into steps that can only be accomplished by following the execution order and the inner relationships, as the stonecutter roughly shapes the block of stone from which the capital is obtained, following a pre-determined order of gestures.[18] The mathematical representation demands similar procedures, which cannot be learned from software manuals; they can be learned through the study of descriptive geometry, of its history and its applications.[19]

Last, but not least, is the problem of the paradigmatic and syntactic chaos that reigns today in the applications of digital representation. In fact, each of these applications, even if implementing, substantially, the same well-known algorithms, have different names and the commands are placed in different logical positions and in different hierarchies. First of all, this confusion involves a lot of wasted effort and time in order to switch from one software application to another. It also becomes impossible to rapidly compare the performances of the applications on the market. Worst of all, this confusion leads to the impression that the various CAD applications are quite dissimilar because they apply different theories, whereas, on the contrary, they all use the same methods, the ones we mentioned above.

This attitude of the producers, which evidently responds to market logics and market strategies, will endure until the users start to grant a privilege, as is only right, to those products in which it is easy to recognize that logical order and terminology that the history of descriptive geometry has consolidated, as the one which best responds to the needs of science and of art.

Conclusions

Descriptive geometry is considered by many to be an outdated science, perhaps also because it is confused with Monge's *Géométrie Descriptive*. Having removed this mistaken notion, there is still another, namely, that descriptive geometry is the science that teaches how to represent objects of three dimensions on a two-dimensional support. Descriptive geometry has this capability too, but it is only one among several. Indeed, descriptive geometry is, first of all, the science that teaches to construct shapes of three dimensions, by means of a graphic solution that simultaneously controls the metric, formal and perceptive aspects. If we agree on this definition, we can accept the idea that this science is still useful and that it is open to further development.

In the preceding pages, I tried to show that this renewal of the ancient science is possible. It is possible to augment the graphic methods based on central and parallel projections, adding the methods currently used in the digital representation, namely mathematical representation and numerical representation. It is possible and useful to develop the number and the quality of the geometric tools used in construction processes, from the straight line to the circle (straightedge and compasses), to the conics and the quadric surfaces. It is possible and useful to take advantage of the synergy between the synthesis of the images and the analysis of the calculation (as Monge already hoped for), introducing into the construction processes geometric loci whose use, in the past, was only hypothesized, such as, for instance, the barycentre of a solid. It is possible to reassess the wide field of the applications of descriptive geometry, obtaining innovative results, like interactive dynamic perspective. It is possible, and necessary, to normalize the paradigm and the syntagm of the terms that are used in the digital applications, so that it would no longer be as demanding as it is today to change from one system to another and thus make the most of the capabilities of each of them.

Most of all: it is possible to give descriptive geometry a future and to give the digital applications the dignity of the noble history that belongs to them.

Notes

1. See Théodore Olivier, *Additions au Cours de Géométrie Descriptive*, Paris, 1847, Préface, XV.
2. See [Chasles 1837: Note XXIII, Sur l'origine et le developpement de la Géométrie descriptive].
3. In all, there exist seven manuscripts of Piero's treatise, which are kept, respectively, in Parma, Reggio Emilia, Milan (two), Bordeaux, London and Paris. The most famous printed edition is the one edited by Giusta Nicco Fasola, which appeared in two editions, in 1942 and in 1984. The Foundation Piero della Francesca is, at the present time, working on the National Critical Edition. See [Besomi, Dalai Emiliani and Maccagni 2009].
4. The term is in quotes because it does not come from Piero's language, even if it belongs to his geometrical conception. Here are meant the lines, perpendicular to the ground line, which pass through the projections, first and second, of a point in space. In Italian, *linee di richiamo* literally translates the expression *ligne de rappel* of the French School and refers to the use that is made of these lines in the constructions.
5. See [Monge 1798: Programme, p. 2].

6. After the publication of the *Mathematical Collections* written by Pappus of Alexandria, translated from Greek by Federico Commandino in 1588, François Viète challenged the mathematicians of the time to give a solution of the Apollonian problem, enunciated in the *Collections*: given three geometric entities chosen among points, straight lines or circles, construct the circumference (or the circumferences) that touches all of them. In 1596 Adriaan Van Roomen published a solution that used conics to construct the centres of the circumferences that met with the conditions posed by the statement. Viète replied in 1600, criticizing, in a very sarcastic tone, Van Roomen's solution. The reason for Viète's irony is in the fact that Van Roomen had failed to observe the rule which imposes the use, in the constructions, of solely the circle and the straight line. See [Viète 1600].
7. See [Mascheroni 1797: Prefazione, VII].
8. Ample evidence of Coxeter's way of working with models of any kind, including physical models, can be found in [Roberts 2006].
9. These denominations are commonly used within the ambit of information technology, but were only recently introduced into the field of descriptive geometry. See [Migliari 2009].
10. In order to avoid useless complications, I am intentionally not considering other methods that may seem to be a hybrid between these two, like the Subdivision Surfaces.
11. See [Gaultier de Tours 1812]. Gaultier's *Mémoire* has been recently analysed and discussed in [Fallavollita 2008].
12. See [Migliari 2008a, 2008b].
13. The procedure, simple and effective, has been found by Marta Salvatore during her studies on the prodromes of descriptive geometry in Amédée François Frézier. See [Salvatore 2008a, 2011].
14. See the catalogue of *Raccolte Museali Italiane di Modelli* [Palladino n.d.].
15. See [Migliari 2008c].
16. See [Migliari 2005a, 2005b].
17. This is the path followed, for instance, by Gino Fano in his *Lezioni di Geometri Descrittiva* [1925]. The procedure is simple and elegant in its graphic realization, but nearly unfeasible in the mathematical representation. In fact, the presence of functions that cannot be calculated exactly, like the square root of five, induces errors which, even if very small, are bigger than the tolerances of the most advanced systems. On the contrary, procedures that simulate the physical construction of the solid, like the one suggested by Gino Loria in his *Metodi Matematici* [1935], are very effective even in the digital field.
18. N. Asgari [1988] has conducted important studies on the algorithms for the working process, in a marble quarry, of the Corinthian capital. On this same topic see also Marco Greco's doctoral dissertation [1996].
19. The application of mathematical algorithms to the study of descriptive geometry and its history has produced many results of remarkable interest. Among these, I would like to recall Camillo Trevisan's studies on stereotomy, on the perspective of the ancients and on axonometry in the nineteenth century [2000a, 2000b, 2005].

References

ASGARI, N. 1988. The stage of Workmanship of the Corinthian Capital in Preconnesus and its export form. Pp. 115-125 in *Classical Marbles: Geochemistry, Technology, Trade*, M. Herz and M. Waelkens, eds. Dordrecht: Kluwer.

BESOMI, O., M. DALAI EMILIANI and C. MACCAGNI. 2009. I lavori in corso per l'edizione del De Prospectiva Pingendi. *1492, Rivista della Fondazione Piero della Francesca* II, 2: pp. 105-106.

CHASLES, Michel. 1837. *Aperçue historique sur l'origine et le développement des méthodes en géométrie, Particulièrement de celles qui se rapportent a la Géométrie moderne, suivi d'un, Mémoire de Géométrie sur deux principes généraux de la Science, la Dualité e l'Homographie.* Brussels: M. Hayez, Imprimeur de l'Académie Royale.

FALLAVOLLITA, Federico. 2008. L'estensione del problema di Apollonio nello spazio e l'École Polytechnique. In *Ikhnos, Analisi grafica e storia della rappresentazione.* Siracusa: Lombardi editori.

FANO, Gino. 1925. *Lezioni di Geometria Descrittiva, date nel R. Politecnico di Torino*. Torino: Parvia.

GAULTIER DE TOURS, L. 1812. Mémoire, Sur les Moyens généraux de construire graphiquement un Cercle déterminé par trois conditions, et une Sphère déterminé par quatre conditions; Lu à la première Classe de l'Institut, le 15 Juin 1812. *Journal de l'école polytechnique* XVI: 124-214.

GRECO, Marco. 1996. Gli algoritmi del capitello corinzio, Procedure di tipo algoritmico nel disegno, il rilievo, la realizzazione del capitello corinzio. Ph.D. dissertation, Rilievo e Rappresentazione dell'Architettura e dell'Ambiente, Rome.

LORIA, Gino. 1935. *Metodi Matematici, Essenza – Tecniche – Applicazioni*. Milan: Hoepli.

MASCHERONI, Lorenzo. 1797. *Geometria del compasso*. Pavia: Presso gli Eredi di Pietro Galeazzi.

MIGLIARI, Riccardo. 2009. *Geometria descrittiva. Volume I – Metodi e costruzioni*. Novara: De Agostini Scuola.

———. 2008a. Rappresentazione come sperimentazione. In *Ikhnos, Analisi grafica e storia della rappresentazione*. Siracusa: Lombardi editori.

———. 2008b. Il problema di Apollonio e la Geometria Descrittiva – The Apollonian problem and Descriptive Geometry. *Disegnare, idee immagini – ideas images* 36: 22-33.

———, ed. 2008c. *Prospettiva dinamica interattiva, La tecnologia dei videogiochi per l'esplorazione di modelli 3D di architettura*. Rome: Edizioni Kappa.

———. 2005a. La prospettiva e Panofsky, Panofsky and Perspective. In *Disegnare, idee immagini – ideas images* 31: 28-43.

———. 2005b. Has perspective a future? (Has Man a future?). *Ikhnos, Analisi grafica e storia della rappresentazione*. Siracusa: Lombardi editori.

MONGE, Gaspard. 1799. *Géométrie Descriptive, Leçons données aux Écoles Normales. L'an 3 de la République*. Paris : Baudouin, Imprimeur du Corps législatif de l'Institut national.

OLIVIER, Théodore. 1847. *Additions au Cours de Géométrie Descriptive, Démonstration nouvelle des Propriétés Principales de Sections Coniques*. Paris: Carilian-Goeury et Vor Dalmont Éditeurs.

PALLADINO, Nicla, ed. n.d. Raccolte Museali Italiane di Modelli per lo studio delle Matematiche Superiori. http://www.dma.unina.it/~nicla.palladino/catalogo/.

PIERO DELLA FRANCESCA. 1984. *De prospectiva pingendi*. G. Nicco Fasola, ed. Florence: Le Lettere.

ROBERTS, Siobhan. 2006. *King of Infinite Space: Donald Coxeter, the Man Who Saved Geometry*. New York: Walker Publishing Company.

SALVATORE, Marta. 2012. Prodromes of Descriptive Geometry in the *Traité de stéréotomie* by Amédée François Frézier. *Nexus Network Journal* 13, 3: 671-699.

———. 2009. Intersezioni piane tra superfici quadriche. Pp. 280-295 in Riccardo Migliari, *Geometria descrittiva*, vol II. Novara: De Agostini Scuola

———. 2008a. La stereotomia scientifica in Amédée François Frézier, Prodromi della geometria descrittiva nella scienza del taglio delle pietre. Ph. D. dissertation, Scuola Nazionale di Dottorato in Scienze della Rappresentazione e del Rilievo, Florence.

———. 2008. Contributi alla ricerca delle sezioni circolari in un cono quadrico. Pp. 401-406 in *La geometria tra didattica e ricerca*, Barbara Aterini and Roberto Corazzi, eds. Florence: Dipartimento di Progettazione dell'Architettura dell'Università degli Studi di Firenze.

TREVISAN, Camillo. 2005. Sull'uso delle assonometrie oblique generiche nella rappresentazione dell'architettura. *Disegnare Idee e Immagini – ideas images* 30: 66-71.

———. 2000a. Sulla stereotomia, il CAD e le varie trompe d'Anet. Pp. 27-53 in *Geometria e Architettura, Strumenti del dottorato di ricerca in Rilievo e Rappresentazione*, Riccardo Migliari, ed. Rome: Gangemi editore.

———. 2000b. La prospettiva degli Antichi nella costruzione proposta da Erwin Panofsky. Analisi e confronto sinottico. *Disegnare Idee e Immagini – ideas images* 17: 59-64.

VAN ROOMEN, Adriaan. 1596. *Problema Apolloniacum, Adrianum romanum constructum*. Wirceburgi: Typis Georgij Fleifchmanni.

VIÈTE, F. 1600. *Apollonius Gallus seu, Exsuscitata Apollonii Pergaei peri epafwn Geometria, Ad V. C. A. R. Belgam*. Paris.

About the author

Riccardo Migliari was born in 1947 and graduated with a degree in Architecture in 1972. He is University Professor in Fundamentals and Applications of Descriptive Geometry since 1990 at the Faculty of Architecture at the 'La Sapienza' University in Rome. His teaching and researching career started right after university graduation and has continued uninterruptedly. He has taught at the Faculties of Architecture 'G. D'Annunzio' in Chieti and at the 'La Sapienza' and the 'Terza Università' in Rome. He also taught Rélevé Instrumental et Photogrammetrie Architecturale at the Post-graduate School of the École Polytechnique d'Architecture et Urbanisme in Algeria, within the International Co-operation Agreement. From 1995 to 2002 he coordinated the Post-graduate School in Architectural and Environmental Survey and Representation and directed the Laboratory of Close Range Photogrammetry at the Department of Representation and Instrumental Survey and a wide range of research activities carried out within this department. He is assiduously engaged in research, particularly in the areas of descriptive geometry and of representation and instrumental survey of architecture. He directed, as Scientific Manager, the architectural survey of the Coliseum in Rome during the preliminary studies for the restoration of the monument undertaken by the Archaeological Superintendence of Rome. From 2003 he has dealt in particular with the renewal of the studies on the scientific representation of space, within in the evolutionary picture of the descriptive geometry, from the projective theory to the digital theory and from the graphical applications to the digital modelling. Since 2008 he has coordinated the "National Laboratory for the Renewal of Descriptive Geometry". He is the author of approximately ninety publications, many of which are monographic. Some of his works can be found at http://riccardo.migliari.it and at http://w3.uniroma1.it/riccardomigliari/Ref/ Default.aspx.

José Pedro Sousa

Faculty of Architecture
University of Porto
Via Panorâmica S/N
4150-755 Porto, PORTUGAL
mail@jpsousa.net

Keywords: digital fabrication,
CAD/CAM software, computer
modelling, computer numerical
control milling, CNC

Didactics

Material Customization: Digital Fabrication Workshop at ISCTE/IUL

Abstract. This paper describes the workshop which took place during the international symposium "Digital Fabrication – a State of Art", which took place at the School of Technology and Architecture ISCTE-IUL, on 15-16September 2011. Its main goal was to introduce a group of about twenty people to the use of digital fabrication in architecture, in a country where these technologies are not yet fully implemented in architecture schools and curricula.

This paper describes the workshop held during the international symposium "Digital Fabrication – a State of Art", which took place at the School of Technology and Architecture ISCTE-IUL, on 15-16September 2011. Its main goal was to introduce a group of about twenty people to the use of digital fabrication in architecture, in a country where these technologies are not yet fully implemented in architecture schools and curricula. Thus, the workshop gathered an international, heterogeneous group of participants, ranging from students with no experience in 3D modeling to young professionals with skills in parametric design with Grasshopper. For this initiative, the school invited three Portuguese companies to support it. At the technological level, Ouplan installed a 3-axis CNC milling machine. At the material level, Amorim provided pure cork agglomerate (*black*) and compound cork agglomerate (*white*) boards, while Investwood supplied Valchromat panels in different colors. The timeframe for the workshop was very short. In total, two afternoons sessions were planned to offer the participants an experience of a continuous process of digital design (CAD) and fabrication (CAM). All these human, technological, material and time parameters defined the workshop conditions (fig. 1).

The starting point consisted in a brief introduction to the use of digital design and manufacturing technologies in architecture. An increasing geometric freedom and the possibility for engaging serial variation logics of production were highlighted as two relevant digital opportunities that have been shaping many of the most relevant contemporary buildings.[1] In relation with the workshop, the use of CNC milling fabrication was presented in detail as a very versatile process for material customization in architecture. Through such technology, it is possible to execute both 2D (e.g., contour cutting) and 3D subtractive fabrication operations (e.g., surface milling) through the moving action of a spindle tool. Complementarily, the nature and properties of the testing materials were described and illustrated. As a brief remark, cork products are lightweight, ecological materials that can perform several functions at once. Known as good thermal and acoustic insulation material, pure cork granulate is a 100% natural material. It results from a thermal process that provokes the volume expansion and self-agglomeration of the granules against the walls of an autoclave without any artificial bonding substance. Overcoming its conventional hidden application as a wall insulation material, it has started being used as an exterior material for building facades since Álvaro Siza built the Portuguese Pavilion at the EXPO 2000 in Hannover. From then on, many buildings have explored such possibilities with success.[2] In contrast, compound cork implies the use of adhesives to bond together the small triturated granules.

DOI 10.1007/s00004-012-0123-7; *published online* 21 September 2012

Fig. 1. The workshop environment. Above, images from the studio where the design part took place. Below, images of the available materials (pure cork agglomerate, compound cork agglomerate and Valchromat), and of the fabrication area with the 3-axis CNC milling machine. Both spaces were next to each other, which promoted the interaction between the design and fabrication phases

Due to its properties and cost, its conventional application is mostly related to interior and product design. Valchromat can be considered as an evolution of medium-density fibreboard (MDF) material, which integrates color pigments throughout its fiber composition. Available in different colors and with a high density, these are highly attractive materials for interior design and furniture applications.

Within this context, three factors were pointed out as the most influential ones for the production of any customized element through CNC milling processes:

- the design geometry
- the machining instructions
- the material properties

Occurring in the CAD software, the first one sets the overall design geometry (e.g., form, contour, perforations, etc.) of the desired piece by means of 2D curves and 3D surfaces. The second factor is concerned with programming the machining instructions with the CAM software. The definition of fabrication parameters like the mill tool and tool path geometries, and the step-over and cutting-depth values, have a profound impact in the final surface texture quality. Finally, the properties of a material composition can add extra qualities to the piece. For instance, by milling layered materials (e.g., plywood) the differences of their variable composition in thickness will show up at the final surface, producing frequently unforeseen and non-designed effects. It is the conscious negotiation of these three factors that can drive architects to achieve a specific material effect by finding an efficient and feasible CAD/CAM strategy to produce it. These multi-level

intricacies and their design potentiality can be learned from the theoretical investigations and pioneering CAD/CAM projects developed by Bernard Cache and Patrick Beaucé since the mid-1990s.[3]

Given these conditions, the challenge for the workshop was launched: the design and fabrication of a customized 50x50 cm panel by groups of two people. The first step was the selection and definition of the working material out of the three available. The groups could decide between preparing homogeneous blocks made out of layers of a unique material or, instead, heterogeneous blocks made out by gluing different material boards. Then, at the design level, it was proposed to explore a surface generation strategy combining the use of an image editor (e.g., Photoshop) with Rhinoceros. By producing a grayscale picture, the "Heightfield from Image" command in Rhinoceros enables the automatic generation of 3D digital surfaces by assigning to each grayscale value a specific height on the z-axis. This methodology proved to be very intuitive and powerful for two reasons. On the one hand, it allowed CAD beginners to generate interesting forms very quickly. On the other hand, by understanding the logics of graphic-to-surface translation, the careful manipulation of the picture can generate highly sophisticated geometries in a controlled way, which could be very time-consuming to achieve by means of other conventional modeling techniques (e.g., loft, sweep...) (figs. 2 and 3).

Fig. 2. A few examples of the automatic generation of 3D surfaces in Rhinoceros from 2D grayscale pictures. The logic behind this process can be easily understood and its potential of application is very large

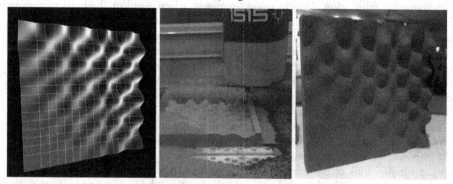

Fig. 3. A custom panel in compound cork, whose design was conceived by using the "Heightfield from Image" function in Rhinoceros

More experienced participants were allowed to explore alternative design modeling strategies. In one of the works, to control the modeling of a Voronoi pattern for milling, the use of Grasshopper was important to quickly and automatically generate a diagram out of a point cloud (fig. 4).

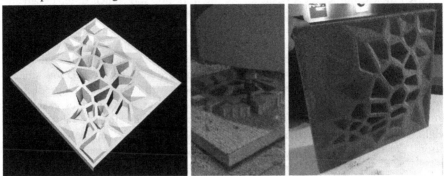

Fig. 4. A custom panel in compound cork, whose design was conceived in Rhinoceros with the help of Grasshopper to control the generation of a Voronoi diagram

During the design process, the groups were encouraged to digitally preview the final result of their intentions by simulating the final aspect of their project. This was tested either through 3D modeling operations (e.g. subtraction of solids and surfaces) or through machining simulation in the CAM software. By doing this, the groups could start sensing how the material consequences at the fabrication level could affect the design intention in a retroactive way. After digitally previewing the effects, some of them went back to adjust their design geometry to achieve better results. Unlike traditional linear processes, these feedback loops between design and fabrication are a major opportunity for creativity arising from the CAD/CAM environment. If there had been more time available for the workshop, the fabrication of some parts would have been desirable.

To allow every group to have an experience of digital fabrication, a maximum machining time of thirty minutes was assigned to each. As soon as the first ones got their design project ready, the CNC machine started working. The technician from Ouplan took care of the CAM programming and the material set up on the machine, following the participants machining interests. Almost all works were finished within the assigned timeframe for fabrication. Due to some geometric delicacy and/or the need to remove large amounts of material, some works required more time to be finished. Nevertheless, such incompleteness was not a problem at all, as the successful materialization of the design intention was proved with the partial fabricated part. These same works were also important to raise some critical discussion about the use of CNC milling machine technologies in architecture. Unlike traditional manual or mechanical processes, the key cost issue in digital fabrication is not so much dictated by the geometric complexity of the work but, instead, by the fabrication time consumed. These digital conditions set up a new paradigm shift in the logics of production. For instance, when trying to fit the project in the budget, architects may adjust the fabrication parameters (e.g., increasing the step-over values) to achieve a faster fabrication time and consequently, a reduction of cost. Such decisions at the fabrication level prevent them from quitting or simplifying their design geometry in the CAD file and can sometimes bring surprising effects to the project.

An overall look at the panels produced in the workshop can make evident two different tendencies (fig. 5).

Fig. 5. Examples of panels that demonstrate the influence of the material properties in their final expression. On the left, the colored layers of the Valchromat sandwich add a kind of topographic effect to the intricate perforations. On the right, a custom made sandwich of compound and pure cork agglomerates showing transitions in material color that are not "designed" in CAD files

On the one hand, there are works where fabrication was simply a way to translate a CAD geometry into a physical element. In this case, fabrication parameters and the homogeneity of the raw material did not influence the final result. Indeed, everything was thought, decided and prescribed in the CAD file. On the other hand, there are others that reveal a more holistic approach to digital production. In these works, it is clear how the three factors described above (i.e., design, machining and material parameters) were explored together, to conceive and produce a specific material effect in the end (fig. 6).

Fig. 6. An overview of the panels produced during this short workshop on digital fabrication

The impact of this workshop initiative can eventually be understood at three levels. First, together with the Symposium talks, it was an effort to help the organizers in motivating the academic community to adopt such technologies in the school. At the same time, the experience allowed the participants not only to acquire practical skills in CNC milling processes, but also to develop a critical understanding about the impact of digital fabrication at the formal, material and methodological levels in architecture. For instance, it clearly showed the importance of exploring feedback loops between design and fabrication information to support creative decisions, and the possibility to engage with non-standard modes of production in architecture. Finally, the workshop could be seen as an experience for industrial inspiration. The work produced can serve as an invitation for the material companies that supported the workshop to incorporate a higher degree of customization in their commercial products and production processes by means of digital fabrication technologies.

Acknowledgments

The author wants to thank Luís de Sousa for his precious help in assisting the students, Rui Salgueiro from Ouplan for his tireless and efficient collaboration with the CNC milling work, to Amorim and Investwood for supporting this workshop with their interesting materials. A final word to thank the organizers for the invitation and support, and to the participants for their curiosity and efforts during the workshop.

Notes

1. Branko Kolarevic's *Architecture in the Digital age. Design and Manufacturing* [2003] was mentioned as a clear source of references about the innovative aspects arising from the use of digital technologies in architectural practice.

2. A detailed analysis of cork and the emergent digital possibilities for its application in architecture can be found in my Ph.D. dissertation [Sousa 2010].

3. Exploring the use of CAD/CAM technologies, Bernad Cache and his partner Patrice Beaucé have produced, among other projects, several wooden panels with customized surface and texture effects driven by geometric principles. Their book *Objectile* [2007] shows some of these works and presents an important critical reflection about the possibility for digital variation in the design and manufacturing of architecture.

References

BEAUCÉ, Patrick and Bernard CACHE. 2007. *Objectile. Fast-Wood: A Brouillon Project.* New York: SpringerWien.

KOLAREVIC, Branko. 2003. *Architecture in the Digital Age: Design and Manufacturing.* New York: Spon Press.

SOUSA, José Pedro. 2010. From Digital to Material. Rethinking Cork in Architecture through the use of CAD/CAM technologies. Ph.D. dissertation. Instituto Superior Técnico, Technical University of Lisbon.

About the author

José Pedro Sousa is Professor at FAUP (Porto) and a Guest Professor at dARQ/FCTUC (Coimbra), teaching courses on geometry, computation and digital project. After graduating in Architecture at the FAUP (Porto), he earned a Master degree in Genetic Architectures from the ESARQ-UIC (Barcelona) and a Ph.D. in Architecture from the IST-UTL (Lisbon). Interested in exploring new conceptual and material opportunities emerging from the use of computational design and fabrication technologies, he has developed an intense activity merging research, teaching and design practice in architecture. More information can be found on his website, www.jpsousa.net.

Book Review

Mark A. Peterson

Galileo's Muse: Renaissance Mathematics and the Arts

Cambridge, MA: Harvard University Press, 2011

Reviewed by Kim Williams

Corso Regina Margherita, 72
10153 Turin (Torino) Italy
kwb@kimwilliamsbooks.com

Keywords: Mark Peterson,
Galileo, history of science,
Dante, Divine Comedy,
geometry, Renaissance

Galileo Galilei (1564-1642) was one of the universal thinkers of the late Renaissance who was at home in both of the "two cultures", the sciences and the humanities, although today we think of him almost exclusively as a scientist, and in a still more restricted sense, as an astronomer. Mark Peterson's recent study, *Galileo's Muse*, does much to broaden our conception of the man known as the "the Father of modern science". This is no easy task, as Peterson notes:

> Galileo's mathematics is just as essential to his humanism as is his erudition in the arts, but studying these things in combination requires more kinds of expertise than perhaps anyone can honestly claim (p. 6).

One kind of expertise that he drew on in painting a new picture of Galileo was a knowledge of Dante (1265-1321 ca) and some of the imagery used in his *Divine Comedy*. Dante is convincingly presented as a precursor of Galileo.

The book begins with an introduction to Galileo himself, his background and especially his education: two years of formal education at a monastery, where he probably mastered Aristotelian logic; he also became well-versed in Latin and, at the knee of his father, Vincenzo Galilei, music theory. Then, when he was 19, came a life-changing encounter with Euclidian geometry, studied under the guidance of Ostilio Ricci (1540-1603), court mathematician to Medici Grand Duke Francesco I. Along the way he also became accomplished in drawing, painting and poetry: a true Renaissance man.

But what was mathematics to Galileo? According to Peterson, "What Galileo overheard from Ostilio Ricci was just ... an abstraction" (p. 25), one whose meaning was unclear. What Galileo did over the course of his life was to try to make that meaning explicit and crystal clear.

Some of the ways he did this are evidently drawn from classical sources. What Peterson argues is that Galileo's most important sources were not the works and authors of the two major periods that preceded Galileo's own – the Roman and the Medieval– but of classical Greece (with one significant exception, Dante). The Renaissance was the period that brought many classical treatises back to light, but the science of Galileo's day

– indeed, Galileo himself – was pushing the envelope, giving mathematics new applications and new meanings that went beyond where the Greeks had taken it. For example, Peterson lists four of the possible meanings of geometry when Galileo was studying it: 1) an intellectual ornament for the nobility and those frequenting the nobility; 2) a body of finite knowledge perfected in an earlier epoch, rediscovered and transmitted, to be studied, understood and especially safeguarded, but not enlarged upon; 3) a practical tool in the day-to-day business of commerce, surveying, construction and so forth; 4) the geometry of astronomy, in the context of the Quadrivium, taught in the university. One of Galileo's achievements was to add a fifth meaning to that list: geometry as a metaphor for nature, a tool for understanding the world around us.

One of the very interesting aspects of *Galileo's Muse* is that it paints a very good picture of how science, including mathematics, is *done*: coming across a problem by reading, hearing, picking up or observing, then thinking, reasoning and insight, then experiment, then proof.

A clear, concrete example is Galileo's discovery of and work with the scaling laws, one of his best known results (a more detailed examination of Galileo's scaling laws than the one presented in *Galileo's Muse* is given in [Peterson 2002]). He began working with scale in a reflection on the dimensions of Hell as depicted by Dante. Seeking Medici patronage, in 1588 a young Galileo presented two lectures comparing rival architectural models of Dante's description in the *Inferno*. One of the key questions was whether the roof in the scheme conceived by Florentine architect Antonio Manetti (the one the audience of the lectures was rooting for) would be strong enough to support itself. In his lecture Galileo maintained that it would, a provided an explanation based on dimensions and scaling to support that. Sometime later, however, he became aware of a ghastly error in his reasoning: the model he had upheld as correct was actually fatally flawed. The reason lay precisely in scaling:

> Galileo had described a roof like a dome thirty braccia wide and four braccia thick … as evidence that a roof 3,000 miles wide and 400 miles thick would be strong enough. The small scale model *would* be strong enough, but just barely. … Scaling it up by one braccia to 100 miles is a factor of 300,000. As Galileo had now realized, the scaled-up dome of Manetti's Inferno would be weaker by that same enormous factor and would instantly fall… (p. 228).

It isn't known exactly when Galileo discovered his error, as he must have taken great pains to keep it a secret. But he did finally publish his correct scaling laws fifty later, in his *Two New Sciences* published in 1638, his last book and just four years before his death. There the scaling laws are presented in relation to the arts of shipbuilding (large ships out of water are liable to break under their own weight, where smaller ships do not) and architecture (a wooden beam's strength in resisting moment depends on the proportion of the transverse force to the cube of the beam's diameter). The famous explanation given to illustrate the scaling laws clearly and explicitly is that of why it is impossible for there to be a giant man, based on how large the bones would have to be:

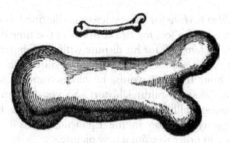

In *Two New Sciences*, written in dialogue form to engage the reader and allow Galileo to use a Socratic method in expounding his discoveries, we have in his own words all the steps in doing science. Coming across a problem: "At times also I have been put to confusion and driven to despair of ever explaining something for which I could not account, but which my senses told me to be true" [Galileo 1914: 1]. Thinking, reasoning, insight: "My mind, like a cloud momentarily illuminated by a lightning-flash, is for an instant filled with an unusual light, which now beckons to me and which now suddenly mingles and obscures strange, crude ideas" [Galileo 1914: 3]. Trial and error: "… if we take a wooden rod of a certain size and length, fitted, say, into a wall at right angles, i.e., parallel to the horizon, it may be reduced to such a length that it will just support itself, so that if a hair's breadth is added to its length it will break under its own weight" [Galileo 1914: 4]. Proof: "…we can demonstrate by geometry" [Galileo 1914: 3].

The valuable addition of Mark Peterson is putting Galileo's well-known discovering of his scaling laws into relation with the little-known lectures on the *Inferno*, allowing us to see how an error led to a correct discovery, in short, how science is done.

The lengthy middle section of the book is devoted to the book's subtitle, *Renaissance Mathematics and the Arts*, and is a discussion of poetry, painting, music and architecture seen through the lens of mathematics. The figures encountered here are well-known: Pacioli, Leonardo, Kepler, Vitruvius. Peterson covers some familiar ground in this discussion of arts and sciences in the Renaissance: classical studies on optics and the discovery of perspective; Platonic solids and golden section; Greek musical theory and the music of the spheres. What is less standard is the inclusion of poetry and specifically Dante. In attempting to depict the universe, the poet was struggling with the two seemingly irreconcilable attributes of God, that of being at once both the edges and the center, qualities described by Robin Evans as envelopment and emanation [Evans 1995: 23]. Here Dante is convincingly depicted by Peterson as building on Aristotle's *On the Heavens* in envisioning Dante and Beatrice in a hypersphere, a device Dante arrived at to depict a finite universe with no edges. Further, Dante's vision of God at the poem's end is based on Archimedes's *On the Measure of the Circle*. Nowadays the study of the *Divina Commedia* is limited to specialized courses in literature. In the Renaissance, however, it was familiar fare, long passages of it were often quoted from memory, and its vivid depiction of the universe taken as genuine cosmology. (In Italy it is still studied in high schools and recently the comedian Roberto Benigni, known to most English speakers for his exuberance at the 1999 Academy Awards, presented a recitation on Italian national television of various canti from the *Divina Commedia* which was absolutely spell-binding, as the high viewer ratings show.) It may have been Dante's exceptional capacity to paint complex mathematical ideas in words that allowed him to provide precisely the meaning of mathematics that the textbooks and treatises did not.

The value of *Galileo's Muse* for this reviewer is the fine job it does of restoring lost knowledge. Knowledge can be lost for two reasons: 1) the time distance; and 2) cultural slant. Galileo is by now "famous" for his dispute with the Church regarding Copernican cosmology, and that single episode has overwhelmed all else that contributes to a well-rounded portrait of him. In other words, he has become a step along the way of a particular narrative of Western cultural history. So yes, he remains known to us, but rather as a two-dimensional billboard that puts all his other aspects in shadow. Except for a brief, almost obligatory mention in the Epilogue, Peterson avoids this episode of Galileo's life altogether in order to paint a new picture.

The final question is, of course, who was Galileo's muse? Of the nine muses, only one concerns herself with a science: Urania, muse of astronomy. Thus Peterson argues that we need a new muse:

"a Muse of Earthly Things, or a Muse of Mathematical Experimental Science. If Astronomy could have a muse in Hellenistic times, then surely these more modern subjects, so aptly associated with Galileo, could have one now. There are many practicing scientists who would be glad to imagine that this Muse was hovering nearby to symbolize to others, or even to themselves, what it is that they do. … Such a divinity would personify Galileo's notion of science with about the right mix of seriousness and lightness…" (p. 297).

He suggests a name; can you guess it?

Galileo's Muse is rather oddly poised between being a scholarly work and a work for a broader readership. The rather general chapters dealing with the arts indicate that the reader was thought to be more familiar with sciences than with the arts (Peterson himself is a physicist and a mathematician). The notes are kept to a minimum, and there is no bibliography, although it is clear that a great deal of secondary literature was consulted, not to mention Galileo's own works. For the layman, a guide to further reading would have been helpful; for the more specialized reader, a bibliography of primary and secondary sources seems in order. This puts the burden on the reader to search for more information, but this is a minor flaw in an otherwise delightful study, one in which even readers with a good background in Renaissance sciences and the arts will find many new insights.

References

PETERSON, Mark A. 2002. Galileo's Discovery of Scaling Laws. *American Journal of Physics* 70: 575-580.

GALILEI, Galileo. 1914. *Dialogues Concerning Two New Sciences*. Henry Crew and Alfonso De Salvio, trans. New York: MacMillan.

EVANS, Robin. 1995. *The Projective Cast: Architecture and Its Three Geometries*. Cambridge, MA: MIT Press.

About the reviewer

Kim Williams is editor-in-chief of the *Nexus Network Journal*.

NEXUS NETWORK JOURNAL · Architecture and Mathematics

Subscription information

ISSN print edition 1590-5896
ISSN electronic edition 1522-4600

Subscription rates

For information on subscription rates please contact:
Springer Customer Service Center GmbH
The Americas (North, South, Central America and the Caribbean)
journals-ny@springer.com
Outside the Americas: subscriptions@springer.com

Orders and inquiries

The Americas (North, South, Central America and the Caribbean)
Springer Customer Service Center
233 Spring Street
New York, NY 10013-1522, USA
Tel.: 800-SPRINGER (777-4643)
Tel.: +1-212-460-1500 (outside US and Canada)
Fax: +1-212-460-1700
e-mail: journals-ny@springer.com

Outside the Americas
via a bookseller or
Springer Customer Service Center GmbH
Haberstrasse 7, 69126 Heidelberg, Germany
Tel.: +49-6221-345-4304
Fax: +49-6221-345-4229
e-mail: subscriptions@springer.com
Business hours: Monday to Friday
8 a.m. to 6 p.m. local time and on German public holidays

Cancellations must be received by September 30 to take effect at the end of the same year.

Changes of address: Allow six weeks for all changes to become effective. All communications should include both old and new addresses (with postal codes) and should be accompanied by a mailing label from a recent issue.

According to § 4 Sect. 3 of the German Postal Services Data Protection Regulations, if a subscriber's address changes, the German Post Office can inform the publisher of the new address even if the subscriber has not submitted a formal application for mail to be forwarded. Subscribers not in agreement with this procedure may send a written complaint to Customer Service Journals, within 14 days of publication of this issue.

Back volumes: Prices are available on request.

Microform editions are available from ProQuest. Further information available at http://www.proquest.co.uk/en-UK/

Electronic edition

An electronic edition of this journal is available at springerlink.com

Advertising

Ms Raina Chandler
Springer, Tiergartenstraße 17
69121 Heidelberg, Germany
Tel.: +49-62 21-4 87 8443
Fax: +49-62 21-4 87 68443
springer.com/wikom
e-mail: raina.chandler@springer.com

Instructions for authors

Instructions for authors can now be found on the journal's website: birkhauser-science.com/NNJ

Production

Springer, Petra Meyer-vom Hagen
Journal Production, Postfach 105280,
69042 Heidelberg, Germany
Fax: +49-6221-487 68239
e-mail: petra.meyervomhagen@springer.com
Typesetter: Scientific Publishing Services (Pvt.) Limited, Chennai, India
Springer is a part of
Springer Science+Business Media
springer.com
Ownership and Copyright
© Kim Williams Books 2012